Laser Beam Profiling: Cost-Effective Solutions

Laser Beam Profiling: Cost-Effective Solutions

Akhil Kallepalli
School of Physics and Astronomy, University of Glasgow, Glasgow, UK

David B James
Cranfield Defence and Security, Cranfield University, Shrivenham, UK

IOP Publishing, Bristol, UK

Multimedia content is available for this book from http://iopscience.iop.org/book/978-0-7503-3835-6.

ISBN 978-0-7503-3835-6 (ebook)
ISBN 978-0-7503-3833-2 (print)
ISBN 978-0-7503-3836-3 (myPrint)
ISBN 978-0-7503-3834-9 (mobi)

DOI 10.1088/978-0-7503-3835-6

Version: 20220401

IOP ebooks

British Library Cataloguing-in-Publication Data: A catalogue record for this book is available from the British Library.

Published by IOP Publishing, wholly owned by The Institute of Physics, London

IOP Publishing, Temple Circus, Temple Way, Bristol, BS1 6HG, UK

US Office: IOP Publishing, Inc., 190 North Independence Mall West, Suite 601, Philadelphia, PA 19106, USA

To Amy, thank you for everything. Our conversations will keep me ever curious and fascinated.

—**AK**

Contents

Preface

In an ideal book, we would be discussing every facet of lasers and their applications, imaging systems and their utilities, and interpretation of results from both laboratory and real-world scenarios. This would include both the history and the ever-progressing scientific discussion, leading to a complete guide for novices and experienced professionals alike. While we are hoping, we may as well hope that the document updates itself every time a breakthrough has been made. However, such a text, although preferable, is a monumental and near-impossible task in a field that changes rapidly. Domains of physics and engineering constantly find new thresholds to surpass and their applications extend into almost every domain from fundamental physics, quantum technologies to popular innovations such as autonomous driving.

Since the first working laser (**L**ight **A**mplification by **S**timulated **E**mission of **R**adiation) was reported in 1960, the laser has on many occasions been referred to as a 'solution looking for a problem.' It has found applications in medicine, military, cosmetic surgery, industry and research, to name a few. Through this text, we provide a brief overview of lasers, a domain of physics graced by the likes of Theodore Maiman and Albert Einstein. Many textbooks available today deal with topics ranging from the quantum physics of stimulated emissions to detailed discussions of each type of laser system available.

We also discuss semiconductor-based detectors. While many silicon-based detectors are available, we focus on CCDs and CMOS sensors. Once again, a brief introduction is given and texts that deal with each of these systems are shared. Although we do not intend to discuss the history and science of lasers and imaging cameras, the literature is intended to be all-inclusive, within its scope.

Laser beam profiling, before and after, interaction with an object or medium could give insight into its nature by observing changes in the beam geometry. The primary objective of this book is to provide undergraduate-level students, academics and laboratory technicians with the knowledge of experimental techniques to use imaging systems for profiling and analysing laser beams. As part of this book, we share downloadable computer programs and templates to assist the reader through the entire methodology. Additionally, video files are also available as a guide (available at http://iopscience.iop.org/book/978-0-7503-3835-6).

Through this book, we hope that the reader is in possession of a refresher guide to lasers and imaging systems and a go-to handbook for monitoring laser beams using imaging sensors.

Author biographies

Akhil Kallepalli

Dr Akhil Kallepalli is a Research Associate in the Optics Group, School of Physics and Astronomy at the University of Glasgow. Having completed a PhD in biomedical optics from Cranfield University, he now works with novel imaging systems and techniques such as single-pixel imaging and its applications. His current research interests include image processing, computational imaging, compressive sensing, biophotonics and remote sensing. His experience includes published research in imaging systems, biophotonics, hyperspectral imaging and remote sensing. He is an active member of multiple professional organisations, passionately contributing with research and outreach activities in different sectors.

Contact information
Website: https://kallepallilab.com/
Address: Room 155, Kelvin Building, University of Glasgow, Glasgow G12 8QQ, UK
Email: Akhil.Kallepalli@glasgow.ac.uk

David B James

Dr David B James is a Senior Lecturer and Head of the Centre for Electronic Warfare, Information and Cyber. He joined the MoD in 1989 and worked on non-linear optical systems and devices, gaining a PhD from the Royal Military College of Science in 2002. In 2008, he joined Cranfield University, conducting research into novel laser applications.

IOP Publishing

Laser Beam Profiling: Cost-Effective Solutions

Akhil Kallepalli and David B James

Chapter 1

Introduction

Abstract

Light acts as a stimulus and induces a multitude of responses from any surface it is incident on. In this chapter, we make a brief overview of the theories of light, lasers and detectors used in optical laboratories around the world. Further, the chapter states the benefits of using imaging sensors for profiling laser beams and provides an overview of the book.

Applications using light for inquiring about physical or biological phenomena are incomplete without an optical radiation detector. In the visible regions, photon detectors have gained a high degree of popularity, given the extent of their utility. These detectors are semiconductor-based electronic components that detect the amount of light 'falling' on their surface and convert it into electrical signals. Optical radiation carries with it energy that is capable of inducing a proportional conversion to electrical signals when incident on the detector material. In this book, we are keen on presenting a method of using silicon-based sensors for beam profiling. The book covers lasers and their working, silicon-based charge-coupled device (CCD) and complementary metal–oxide–semiconductor (CMOS) sensors before combining them for use in the laboratory. The book is in no way an exhaustive thesis on either of these concepts, but attempts to introduce them at an accessible level and provide a guide to using them in the laboratory. The method detailed in the book for using silicon-based imaging detectors as a laser diagnostics tool serves two purposes of power measurement and beam profiling. While laser power is usually measured using a power meter, this method enables the end user to use a single device for quantifying the output of the laser and understanding the geometry of the beam simultaneously. Understanding the beam characteristics and profile is key in many applications where knowledge of the initial and the final measured beam can provide information about the object/medium. This resource includes a video guide on how to calibrate the components needed and to profile the beam itself.

doi:10.1088/978-0-7503-3835-6ch1

1.1 Light

Early scientists such as Pierre Gassendi and Sir Isaac Newton long believed that light was made up of particles called corpuscles. These corpuscles, travelling in straight lines, result in light rays. Until the discovery of light bending around sharp edges (through a phenomenon called diffraction), the particle theory explained most of light's behaviour such as creating shadows, and brightness and darkness due to presence or absence of light. Subsequently, with the discovery of diffraction, light began to be described as a wave. After initially being disregarded, the work of Christiaan Huygens and Augustin Jean Fresnel were proven to show wave behaviour of light. Much later, in the twentieth century, quantum mechanics fused both theories and accepted the dual wave and particle nature of light [1].

Subsequent to Albert Einstein's contributions[1] on the photoelectric effect, he further laid the foundation for understanding stimulated and spontaneous emissions at atomic levels. His paper 'On the quantum theory of radiation' in 1916 was pivotal in the eventual realisation of the first working laser by Theodore Maiman in 1960. Through a deceivingly short article [2] on stimulated emission using ruby, Maiman presented a light amplification device that is one of the most commonly used technologies today. The word *laser* is an acronym for **L**ight **A**mplification by **S**timulated **E**mission of **R**adiation. Although lasers were initially analogous to an *optical* maser, due to its functionality enforcing that of a maser (amplification of microwave radiation) working in the optical wavelength region, this definition was not regarded as the most accurate for the laser [3].

Lasers are varied and widespread, both in terms of type and application. In modern communication systems and audio systems such as compact disc players, the laser is the most important component. Industrial applications utilise the focused power and intensity of the laser for manufacturing, cutting, cleaning and heat treating. Medicine, a domain constantly searching for new and efficient tools, has found lasers to be a handy tool in surgery, whether they be oncology or dermatology. The military utilise it for range finding and targeting, while the supermarkets use lasers at checkout counters [4]. A full list of laser applications would be endless and forever incomplete as with every passing day, research and innovation extends its applications into more and more domains.

1.2 Detectors

Any device that receives and responds to a stimulus can be considered a sensor [5]. When this definition is applied to optical sensors, the analogy unravels to 'an electrical signal as a result of material–light interaction.' Light interacts with objects or media and after phenomena such as reflection, refraction, absorption, interference

[1] Albert Einstein's contributions are fundamental to the concepts covered in this text. Firstly, his work on the photoelectric effect, that won him the Nobel Prize, is the basis for solid-state quantum detectors (photovoltaic and photoconductive devices). These rely on the interaction of individual photons with a crystalline lattice of semiconductor materials. Secondly, his contribution to understanding absorption of energy at atomic levels was the theoretical background for the invention of the laser.

and polarisation before interacting with an optical sensor or detector. Detectors capable of measuring electromagnetic radiation in the ultraviolet to far-infrared are known as light or optical detectors [6]. From a sensor's standpoint, the detected electromagnetic radiation could result in a quantum or thermal response, which effectively defines the broad definition of optical detectors. Quantum detectors function in ultraviolet to mid-infrared regions while thermal detectors function from mid- and far-infrared spectral ranges.

Materials used in detector sensors are sensitive to incident light. The changes in the characteristics of the material cause a change in the current in the detector circuit. The change in current, in turn, results in a change of voltage, creating an output signal. Optical detectors must be chosen with due care and consideration of the application, as the operating wavelengths decide the type of detector and optics[2]. Currently, many detectors are available, capable of capturing ultraviolet, visible and infrared wavelength of the spectrum [7]. The significance of imaging systems operating across all wavelengths of the electromagnetic spectrum is unquestionable, as the number of applications would fill volumes of textbooks. Extensive and exhaustive guides to laser physics or optical detectors are quite the task to formulate. We would like to mention, at this stage, that this is not the motivation for this literature. However, we will share relevant resources through textbooks and pivotal research articles for your reference through footnotes or in dedicated sections for additional resources.

In this text, we provide a basic overview of lasers and silicon-based imaging systems; particularly charge-coupled devices (CCD) and complementary metal–oxide–semiconductor (CMOS) sensors. This overview is to serve as a 'fuel to inspire' introduction to a novice, a reference guide to a laboratory technician setting up laser equipment, and a refresher to experienced researchers and engineers. It will discuss conventional laser power meters before moving on to discuss the benefits of using an imaging system for laser diagnostics. Chapter 2 will provide an introduction to lasers. Thereafter, we will discuss imaging systems and a broad overview of the sensors in chapter 3. A detailed discussion and guide to using imaging systems for laser profiling and measurement (chapter 4) is followed by a description of the potential sources of errors when using this method for laser diagnostics is detailed in chapter 5. Finally, the conclusions presented in chapter 6 complete this work.

1.3 Benefits of using an imaging system

Most laser systems' power is measured using power meters. Conventional power meters, like the Lasermet ADM 1000 (figure 1.1)[3], use integrating spheres combined with photodiode detectors for maximising the degree of light entering and being measured by the detector. However, they do not provide information regarding the beam profile. They are dependable sources of information when measuring the amount of light in a specific direction, either from a laser or from other sources. The

[2] Imaging systems operating in the visible region use glass lenses while thermal cameras, which operate in the long-wave infrared region require germanium or sapphire crystal lenses.

[3] The Lasermet power meter is used in this book to measure the power of the laser, chapter 4.

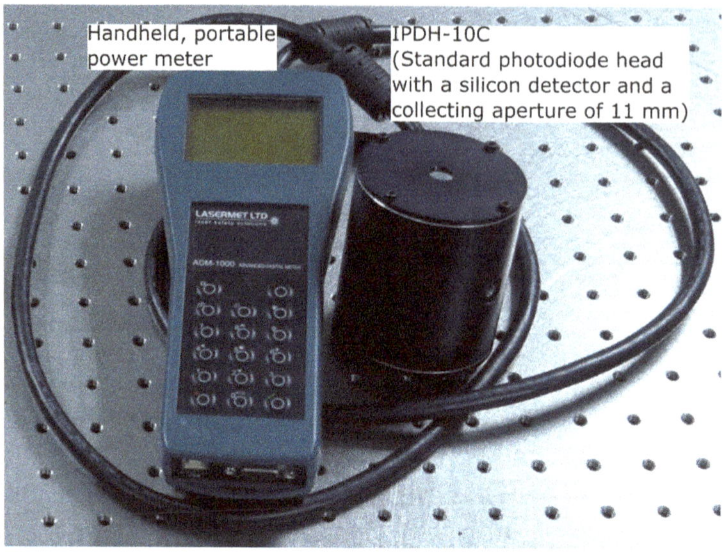

Figure 1.1. The Lasermet ADM1000 offers a handheld accurate power meter that is capable of sampling at 1 MHz and enabling accurate measurement of laser power. The device can be coupled with a range of detector heads such as the IPDH-10C (shown) which offers a response time of 700 ns. The detector head comprises of a collection aperture of 11 mm and a silicon detector that can measure up to 1 W of laser power between 400 and 1100 nm.

digital meter provides a wide range of metrics related to lasers, including peak power, energy per pulse, graphical illustrations and computer interfacing to utilise the data in a multitude of ways. This information is useful in most applications where the beam power and profile are important to monitor.

CCD and/or CMOS sensors can acquire two-dimensional information of the input beam. While the power cannot be measured directly with these sensors, they can measure the footprint and intensity distribution (within the dynamic range of the sensor). Therefore, knowledge of the optical power and/or measuring the beam profile can be combined. This guide provides a step-by-step method of combining beam profiling using imaging sensors while simultaneous monitoring laser power. The advantages of this method are:

- The beam profile information is intuitively available. The method provides an option of both snapshot and continuous monitoring of the beam profile. An analysis of the beam's shape, intensity distribution and profile can be made.
- The laser power can be monitored continuously and simultaneously with the beam profile.
- The consistency of the laser's output can be monitored on a regular basis.
- The distribution of laser hotspots can be mapped using this method. It is important to identify this, as it defines the locations in the beam of the laser where it presents maximum intensity on incidence. This intensity defines the application and utility of the laser.

References

[1] Heintzmann R and Kubitscheck U 2017 Introduction to optics *Fluorescence Microscopy: From Principles to Biological Applications* ed U Kubitsheck 2 edition (New York: Wiley-VCH)
[2] Maiman T H 1960 Stimulated optical radiation in ruby *Nature* **187** 493–10
[3] Allen L 1969 Introduction to the laser *Essentials of Lasers The Commonwealth and International Library: Selected Readings in Physics* ed D Ter Herr (Oxford: Pergamon) pp 3–7
[4] Silfvast W T 2008 Lasers (Module 1.5) *Fundamentals of Photonics* ed A Guenther, L S Pedrotti and C Roychoudhuri (Orlando, FL: SPIE) ch 5 pp 1–44
[5] Fraden J 2010 Dara acquisition *Handbook of Modern Sensors: Physics, Designs and Applications* (New York: Springer) pp 1–12
[6] Fraden J 2010 Light detectors *Handbook of Modern Sensors: Physics, Designs, andApplications* (New York: Springer) pp 461–501
[7] Ready J 2008 Optical Detectors and Human Vision (Module 1.6) *Fundamentals of Photonics* ed C Roychoudhuri (Bellingham, WA: SPIE) p 211

IOP Publishing

Laser Beam Profiling: Cost-Effective Solutions

Akhil Kallepalli and David B James

Chapter 2

Lasers

Abstract

Lasers are devices that can emit coherent light that is focused to a tight spot. These emitted wavelengths along the electromagnetic spectrum and the temporal frequency of emission depends on the excitation medium and the construction of the device. Lasers have evolved to have applications across commercial, medical, industrial, defence and research domains, to name a few. The process of lasing is due to optical amplification and based on stimulated emission of electromagnetic radiation—thus the laser acronym of **L**ight **A**mplification by **S**timulated **E**mission of **R**adiation. In this chapter, we discuss the process of stimulated emission, share an overview of the properties of laser light, introduce the types of lasers and mention laser safety as a key step to handling and working with lasers.

Laser action can generally be described as the behaviour of light absorption and emission by matter. It is an amplifier with positive feedback, i.e. the amplification of emission of photons due to the positive feedback achieved by mirrors. Therefore, materials and conditions where this amplification can be achieved are used to make lasers across different wavelengths, complexities and costs. The repeated cycle of excitation of photons, and these stimulated photons acting as seeds for subsequent excitation results in optical transitions. This contributes to population inversion, where the stimulated emission rate is about the spontaneous emission rate. When the stimulated emission rate exceeds the stimulated absorption rate, an optical gain is achieved. This is a measure of the ability of a laser medium to increase optical power and result in an emitted beam [1].

When the first working laser was created by T H Maiman in 1960, the applications of its ability to produce high intensity and focused light of a specific wavelength were unimaginable. What began initially as a 'solution in search of a problem' [2], lasers are now an important component from research laboratories to industry, medicine to military, audio systems to teaching tools. Maiman reached this breakthrough by

doi:10.1088/978-0-7503-3835-6ch2

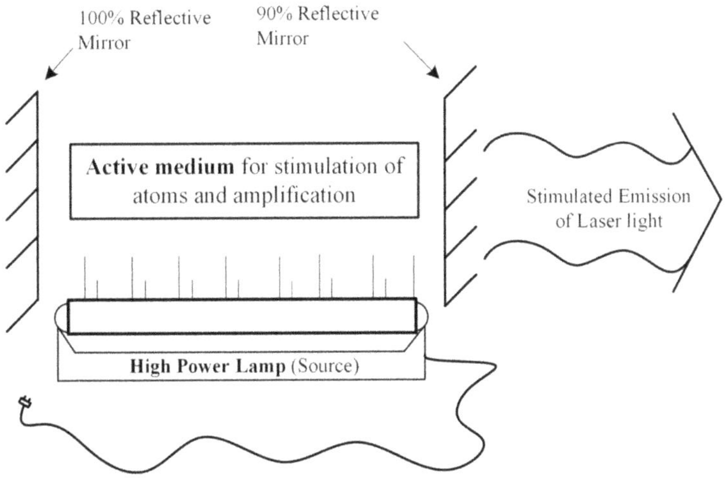

Figure 2.1. A schematic illustration of the components of a laser system, including the source of stimulation, active medium and cavity, and beam output.

illuminating a ruby rod with a high power flash lamp. The ends of the ruby rod were coated with silver to act as the reflective surfaces and achieve amplification [3]. Having already discussed the optical properties of ruby [4] prior to his *Nature* publication, he, in about 300 words [5], described the achievement of population inversion, wavelength emitted, construction of the laser cavity and the discovery that the emitted radiation had a wavelength much shorter than its spontaneous counterpart.

The ideal condition, in which a laser can be realised, is when the atoms can be made to emit at a specific frequency (ν) after interaction with an electromagnetic field of the same frequency. The occurrence of this state ensures that the emissions are not random, but in a definite relationship to the stimulating field—a process known as stimulated emission. A laser consists of four key elements[1], as illustrated in figure 2.1, which are:

1. An active medium—atoms or molecules which can absorb and amplify the light incident on them.
2. An amplifying material—enclosed in a highly reflective cavity tube, it serves the purpose of redirecting initial emissions back into the medium for repeated and increased amplification.
3. A light source—required for replenishing the energy in the active medium being converted to light energy.
4. A beam extraction capability—achieved by using a mirror that allows a certain level of transmission.

A laser consists of two components: a photon amplifier and a positive feedback circuit. The gain spectrum achieved is defined by the frequencies over which the light

[1] Some sources [6] mention three elements by excluding the beam extraction capability and considering it an inclusive characteristic of the amplification cavity.

emission exceeds absorption. Amplification generated by stimulated emission should have a probability of photon emission proportional to the number of photons already present in the medium/system. The positive feedback is achieved by mirrors that return the stimulated emission back into the medium, resulting in optical gain. When this gain exceeds the absorption rate of the medium and photon losses in the resonant cavity, laser action occurs [1].

A functional laser design must concern itself with the gains and losses encountered by the system. Governed by the laws of physics, every stage of the laser system encounters entropy gain and energy loss. When the stimulating light is introduced into the active medium, it excites the electrons to higher energy states. However, they do return to their original states. The mirrors also induce diffraction losses due to their finite dimensions while reflecting the stimulated light through the active medium to create a standing wave [7]. If unable to sustain the necessary strength of electromagnetic field in the cavity, the emission from the system would be short and rather quick. The mirrors forming the cavity reflect the emitted light from the active medium back into it, providing an avenue for achieving feedback. The feedback is key to realising high amplification with a relatively small input. Estimating these losses and accordingly introducing power into the cavity ensures continued emissions at frequency ν and the decaying (loss) of amplitude of the cavity wall will increase once again. In comparison to spontaneous emissions, lasers are known to provide high temporal coherence. This high temporal coherence enables the prediction of the phase and amplitude as the beam travels over large distances. This in turn confirms that lasers have long wave-trains of emissions, thus implying a high degree of monochromaticity [7].

2.1 Light emission, absorption and matter interactions

Einstein's theory of photoelectric effect was regarded as the modern interpretation of light emission and absorption [8]. He postulated that the energy difference of the electron before and after photoejection was the same as that of the incoming photon absorbed, i.e. $h\nu$. The idea of 'quantum jump' was introduced after Einstein's idea was confirmed by its applicability to Bohr's quantum-mechanical model of the atom in 1913. The quantum jump best explains the theory of an electron's transition between two allowed orbits; a concept that is of great significance to stimulated emission.

Quantum jumps[2] occur in three different forms—outer-shell electrons, inner-shell electrons jumping to higher energy levels and outer-electron levels close to the ionisation limit. Each of these require energy. The outer-shell electrons' excitation and subsequent quantum jump is the ordinary form, occurring in the optical region of the spectrum. Inner-shell jumps by the electrons require much higher energies and

[2] Although spontaneous emission is a quantum event, it does not mean that these are rare. At room temperature, spontaneous emission of photons dominates stimulated emission in most circumstances. The LEDs used for general lighting are bright sources of light beams composed almost entirely of spontaneously emitted photons [1].

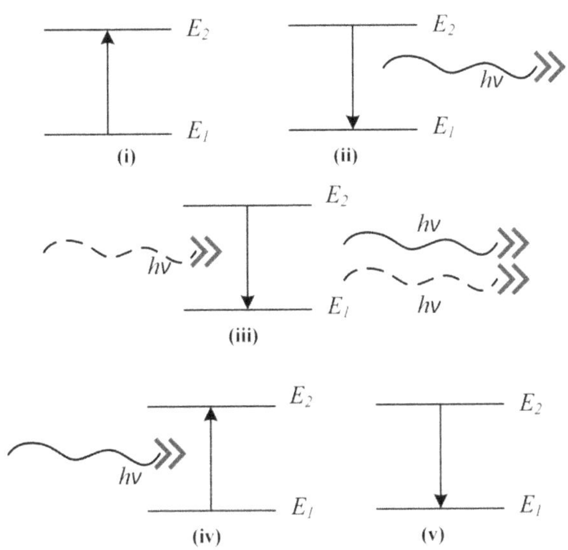

Figure 2.2. Energy conversions between electron levels, illustrating spontaneous and stimulated emissions.

are typically induced by x-ray photons. The electrons in the Rydberg energy levels[3], close to the ionisation limit require a very small amount of incident energy, known to be possessed by infrared and/or microwave photons.

The quantum jump required to induce a stimulated emission cannot be achieved with crystals that are transparent in the optical region. Their transparency is indicative of their nature to not absorb or emit optical photons. However, coloured crystals, like ruby, contain impurities with atoms and molecules that contain electrons ready to make the jump to higher energy levels[4]. This principle is further explained through the energy conversion between two electron shells (figure 2.2). Each atom in the active medium accepts energy through a 'pumping' process, which in turn allows the electron in the inner shell to jump to the higher shell. As the electron returns to its original state, it emits photons. Simultaneously, the atoms also have to deal with the previously emitted photons that are reflected by the mirrors. These prior photons travelling through the active medium are responsible for emission of consequent photons, thereby creating stimulated emissions.

Consider, for simplistic understanding, an atom with two energy levels, E_1 and E_2. The energy of the photon input and output due to the energy conversions is $h\nu$.

[3] Rydberg energy levels are related to the attainment of the ionic state of atoms, with an ionisation energy.
[4] The emission of photons is always accompanied by an electron transitioning from a higher energy state to a lower energy state. The energy emitted is equal to the energy difference of the levels, thereby conserving energy. Maxwell's equations do not account for spontaneous emission because the photon is created from the vacuum or empty state. However, this is to be expected because Maxwell's equations describe the continuous fields resulting from many electrons and photons.

The energy conversion diagrams illustrated in figure 2.2 are discussed, in order, below:

(i) Due to the input energy or pumping process, the electrons in the lower energy shell, E_1, jump to the higher energy shell, E_2, due the absorption of energy ($\Delta E = E_2 - E_1$).

(ii) **Spontaneous emission** of photons (of energy $h\nu = E_2 - E_1$) occurs when the electron returns to its original state, E_1, without any external factors.

(iii) As the photons are reflected back into the active medium of the laser by the mirrors, the photons once again induce a quantum jump. As a consequence of this stimulated energy increase, an identical photon is generated as the excited electron returns to its original state; known as **stimulated emission**.

(iv) Absorption of the photon induces a quantum jump from energy level E_1 to E_2.

(v) In certain cases, the energy transition of the electron from higher to lower level does not produce any radiation. In such cases, the energy is assumed to have been converted to vibrational or rotational energy (for molecules), or translational energy. This is known as **non-radiative de-excitation**

All the energy conversions discussed occur within the active medium of the laser. Lasers are often classified with regards to their pumping process [8] (figure 2.2(i)) and their active medium. In electric-discharge lasers, the pumping process is induced by collision of electrons in a gaseous discharge with the atoms of the active medium, while for optical lasers, the process is initiated due to absorption (figure 2.2(iv)) by a lamp or another laser.

On a quantum scale, an ideal scenario would occur if the lower level of the transition is de-excited rapidly by spontaneous emissions of non-radiative de-excitation. When this occurs, the atom energy level drops to level 1 (E_1) or a lower level prior to the interaction with the incoming photon from a previous emission. If the rate of de-excitation of level 1 is large, the number of atoms remaining at this level is negligible in comparison to level 2 (E_2). This befits the conditions required for lasing action and can be explained in terms of the two 'populations'—the number of atoms with electrons in level 2 (N_2) and the number of photons in the laser cavity (q) [8]. Consistent with our two-level atom consideration thus far, consider the number of atoms with electrons in energy level 1 to be N_1. At thermal equilibrium,

$$\frac{\mathrm{d}N_2}{\mathrm{d}t} = -A_{21}N_2 - B_{21}N_2\rho(\nu) + B_{12}N_1\rho(\nu) = 0, \tag{2.1}$$

where $\rho(\nu)$ is the radiation density associated with the temperature T and A_{21}, B_{21} and B_{12} are Einstein's coefficients. Considering the Einstein's coefficients between the energy levels are equal to B ($= B_{21} = B_{12}$), equation (2.1) evolves to

$$B\rho(\nu)(N_1 - N_2) = A_{21}N_2 \Rightarrow \frac{A_{21}}{B\rho(\nu)} = \frac{N_1 - N_2}{N_2} \tag{2.2}$$

In the case of spontaneous and stimulated emissions, the number of emissions is individually illustrated as

$$n_{\mathrm{sp}} = A_{21}N_2;\ n_{\mathrm{st}} = BN_2\rho(\nu) \tag{2.3}$$

Dividing both emission estimates in equation (2.3), using equation (2.2) and using the Boltzmann distribution for thermal equilibrium,

$$\frac{n_{sp}}{n_{st}} = \frac{A_{21}}{B\rho(\nu)} \Rightarrow \frac{n_{sp}}{n_{st}} = \frac{N_1}{N_2} - 1 = \exp(\frac{h\nu}{kT}) - 1 \tag{2.4}$$

In the optical region of the spectrum, the frequency ν is about 5×10^{14} Hz and the moderate temperature is about 450 K,

$$\exp(\frac{h\nu}{kT}) \approx e^5 \Rightarrow n_{sp} \gg n_{st} \tag{2.5}$$

This is indicative of the higher probability of occurrence of spontaneous emissions in the optical range when the system is in thermal equilibrium. In order to achieve stimulated emissions, the exponential term in equation (2.4) must be $<\ln 2$ so that $n_{sp} < n_{st}$. However, once again, at optical wavelengths (or frequencies) this corresponds to extremely high temperatures. In which case, the number of stimulated emissions only just exceed the number of spontaneous emissions. The primary condition of laser action is that the energy levels should not be in thermal equilibrium and the upper of the energy levels must have higher number of atoms with electrons suitable for the quantum jump. This condition is known as *population inversion*. From equation (2.3), it is evident that for the number of stimulated emissions to be high, the number of atoms (N_2) with electrons at energy level E_2 and the radiation density at frequency ν must be large. With this understanding, it is straightforward to deduce that conditions must be created to achieve population inversion and achieve built up radiation field at frequency ν for laser emission.

2.2 Properties of lasers and laser light

The primary differences between lasers and other light sources are that the laser beams are narrow, illumination is monochromatic and emissions are coherent [6]. These differences are characteristic of laser sources, and along with other properties, require further discussion for a clearer understanding. This understanding serves not only the laser user concerned with its application, but also sheds light on the physics behind the beam properties and light behaviour from stimulated emissions. We will discuss the common parameters used to describe lasers, with the focus on providing the user with the knowledge to understand manufacturer's documentation. In the documentation, you will usually find parameters like central wavelength, divergence, beam dimensions, operating temperature, laser safety class, etc.

2.2.1 Beam width

Laser beams are commonly characterised using their footprint[5]. The divergence is also a key characteristic of laser beams, which can be characterised by profiling the

[5] While the diameter is a common measure to describe laser beam, laser beams can be elliptical as well. Therefore, the beam footprint is described as width in two dimensions. This method covers all laser beam profiles.

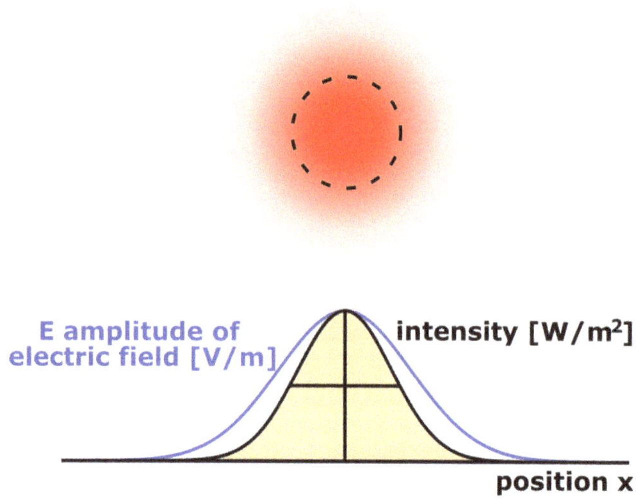

Figure 2.3. Laser beams are normally characterised by their amplitude and intensity. Further, spatially, they are defined by their beam width, diameter and *x–y* profile. Effectively measuring these parameters is described in chapter 4. The dotted line indicates the $1/e^2$ width of the Gaussian beam. This image has been obtained by the authors from the Wikimedia website where it was made available by FDominec under a CC BY-SA 3.0 https://creativecommons.org/licenses/by-sa/3.0/deed.en licence. It is included within this article on that basis. It is attributed to FDominec.

beam at two or more separation distances between the laser module and the detector. The beam width or diameter (figure 2.3) is measured along a specified line that is perpendicular to the beam axis. The definitions of beams are done using multiple methods:

- **Rayleigh beam width**: This measure is defined by the angle between the primary peak of the beam and the first point where no power is radiated (in a particular direction).
- **Full width at half maximum**: Widely regarded as the simplest way to measure the beam profile, this measure is the width of the beam at the point of half intensity with respect to the peak intensity.
- **Knife-edge method**: An early method of measuring the beam width, this method entails slicing a beam with a razor and measuring the power as a function of the position of the razor. The width is defined as the distance between the points where the transmitted power is 10% and 90%, i.e. the distance through which 80% of the beam intensity passes through to the detector.
- $1/e^2$ **width**: This distance is the points along the beam profile between which $1/e^2 = 0.135$ of the power of the laser is measured (figure 2.3).
- **D86 width**: Similar to the previous measures, D86 width is the diameter, centred from the centroid of the beam, of the laser profile that allows 86% of the beam power.

2.2.2 Laser power

One of the primary measurements of the laser beam is the power of the beam. The fact that all the light energy (power measured in Watt (W)) of the laser is contained in a beam and focussed on a small surface area explains the high intensity characteristic of lasers. Laser powers range up to tens of watts or greater for continuous wave lasers, while being much higher for pulsed lasers. Pulsed lasers emit high peak powers over a certain time interval, making their impact much higher. However, on average with respect to time, the laser powers of both types are comparable. To give you an idea, a 1-milliwatt helium–neon laser outputs intensity comparable to the Sun on a clear day. As harmful as the Sun is to the eye over a duration of exposure, lasers also have the same impact. When laser light is incident on the human eye, the lens focusses the light on to a single point on the retina. This focused power increases the intensity of the laser, thereby harming the retina and causing laser damage. However, this focussed delivery of light energy has found applications in medicine, like microsurgery, where surgical lasers focus on the treatment location without affecting the surrounding tissue.

$$\text{Intensity} = \frac{\text{Laser}}{\text{Power area}} \ (\text{W m}^{-2}) \tag{2.6}$$

The relationship (equation (2.6)) between intensity and laser power is also used to estimate the Maximum Permissible Exposure (MPE)[6], a term commonly used in medicine to describe the maximum intensity that skin (or any tissue) can be exposed to without causing any damage.

2.2.3 Monochromatic light and central wavelength

From the perspective of a scientist, one of the most important features of the laser is its ability to deliver monochromatic illumination [9]. Conventional light sources like lamps produce illumination over the visible and near infrared spectrum [6]. It is worth noting that although lasers claim to be able to deliver monochromatic illumination, the ability to produce a single wavelength with absolute 'purity' is practically not achievable. Therefore, through literature, including the current, references to laser systems wavelengths are made with central wavelengths (λ_{CW}) and the range of wavelengths included ($\Delta\lambda$). For example, in the optical region the frequency (ν) is $\approx 5 \times 10^{14}$. While lasers are known to have a $\Delta\nu \approx 1$ Hz, a typical spectroscopically practical laser could have a frequency variation of 100 Hz. Considering, the spectral purity (Q factor[7]) in such cases presents a value of 2×10^{-13}, a practical achievement that is better than most mechanical and electrical resonators by orders of magnitude [8].

[6] MPE for human skin is 2000 W m^{-2} for a 5–10 h exposure period.

[7] Q factor = ($\Delta\nu/\nu$).

The terms 'broadband' and 'narrowband' in this context refer to the spread of wavelengths ($\Delta\lambda$) available within the output of the laser. Broadband lasers have a $\Delta\lambda \approx 300$ nm, while narrowbands have a value in the range of 0.01 nm. If the output of the excitation of the lasing material and subsequent amplification generates an output containing multiple wavelengths, a narrowband filter can be used to block all but one wavelength, thereby achieving the closest approximation of a 'purely' monochromatic light source. This technique is known as *single moding*. Single moding is the closest approximation to absolute monochromatic illumination that can be obtained (where $\Delta\lambda = 0$), and we can only approach this ideal scenario by refining the light sources. Alternatively, the optics used inside the laser cavity, i.e. the reflective mirrors, can be designed to be characteristically reflective only at the wavelength of interest. This method would not allow the other wavelengths in the emission spectrum to reach the threshold of laser action.

2.2.4 Angular properties

From the lowest powered gas lasers used to create scintillating laser displays, to military grade high power lasers capable of destroying incoming missiles, the ability of lasers to hold their intensity over a long distance is one of the advantages of using them as a source. Laser light is a result of multiple reflections between two mirrors in the cavity of the laser system. As an approximation for understanding, light generated within the cavity due to stimulated emissions from the lasing material will oscillate until its energy is greater than the threshold presented by one of the mirrors which is a partial reflector. The distance travelled by the photons before emission from the laser cavity is the product of the length of the laser cavity and the number of times a photon has traversed the length between the mirrors (figure 2.4). Therefore, after a laser photon has travelled this distance, the light waves are eventually planar. Hypothetically, if a laser photon has been reflected over 50 times within the laser cavity, it would appear to have traversed a distance 100 times the length of the cavity with 100 apertures collimating the beam [9].

The directionality of any laser system is expressed as a factor of *full angle beam divergence* (figure 2.4). This is twice the angle of the outer edge and center of the

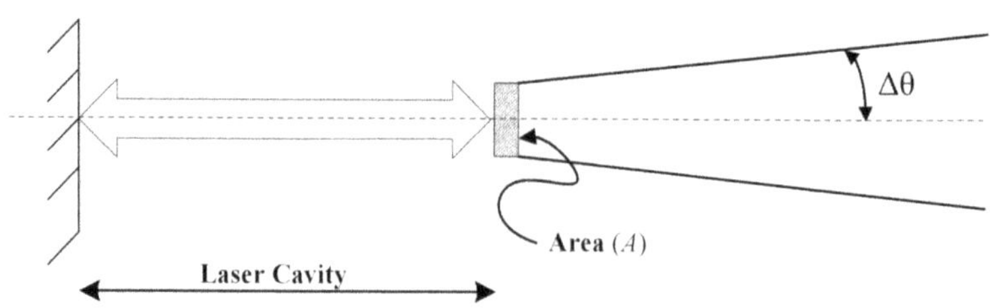

Figure 2.4. Directionality of lasers illustrating the exit aperture of the laser cavity and the divergence of the beam as a function of distance travelled by the beam.

beam, a quantity commonly expressed in radians[8]. Research has shown that this divergence can be reduced even further by using additional optics in the path of the laser beam to focus the light down to a smaller footprint at the expense of loss of intensity[9]. Conversely, a beam's solid angle is defined by the wavelength of light energy and the area of the aperture through which the emission exits the laser cavity. The area of the aperture plays an important role in defining these angular parameters of directionality as it induces diffraction effects to the laser light. The relationship is expressed as below:

$$\Delta\Omega \approx \frac{\lambda^2}{A} \approx (\Delta\theta)^2 \tag{2.7}$$

2.2.5 Temporal and spatial coherence

A measure of temporal coherence for a light source is the time required for two oscillations of frequency differing by $\Delta\nu$ to go out of phase with each other. In other words, coherence is the measure of the light waves' predictability of amplitude at any distance. Like any other light source, lasers also have a defined $\Delta\nu$. However, this value is so low for lasers that they attain extraordinarily high coherence time (the relationship of time to bandwidth is inversely proportional; ($\Delta\tau = 1/\Delta\nu$)). The coherence of the lasers is vital in applications like laser Doppler flowmetry when detecting changes in moving targets of interest as it allows quantification of constructive and destructive interferences through the acquisition of reflected or transmitted light. The coherence length is the distance travelled by the beam, while coherent. Although coherence times can be rather short, light is known to travel long distances in short periods of time. For example, a coherence time of 1 μs still relates to 300 m travelled before the wave-trains of light go out of phase. Coherence measurements are usually performed with instruments known as *interferometers*.

Spatial coherence refers to the area of a wavefront over which the light is coherent [10]. This quantity is unrelated to temporal coherence. A laser source that is temporally coherent, over a distance, is not necessarily always spatially coherent.

2.2.6 Laser speckle

When laser light is incident on a rough surface, the reflection appears to have a textured pattern of a sparkling nature. This phenomenon encountered by laser light is known as laser speckle. Although considered to be a hindrance in some applications, laser speckle offers valuable information regarding the surface of the target. Laser speckle is a manifestation of coherence of the laser, leading to yet another concept of light waves travelling through space—*interference*. Although

[8] Most laser manufacturers refer to the divergence as a unit of milli-radians (mrad). A typical laser system with a 1 mrad divergence has a divergence of 1 mm over a distance of 1 m the beam travels.

[9] When using glass-based optics, there is a loss of 4% energy at each surface due to absorption in the visible region.

beyond the scope of this book to discuss interference at length, a task already achieved best by multiple University-level texts, it is sufficient for the reader to know that the when light waves coincide, they cause patterns of high or low intensity. When the crest of a light waves coincides or 'interferes' with another crest, the intensity doubles. Conversely, when a trough of the wave interferes with a crest, they cancel out to create a low intensity speckle. Although considered as noise in many applications, laser speckle is a measure of the surface structure and roughness and in applications like laser Doppler flowmetry, it represents a measure of blood flow under the skin.

2.2.7 Types of lasers

Lasers are broadly described in terms of their nature (continuous wave, pulsed and ultrafast lasers) and lasing medium. While describing lasers, the reference to the output nature of the laser and the lasing medium are inter-related as the output generated is directly dependent on the active medium in the laser cavity. The relationship between lasing medium and the wavelength output is illustrated in figure 2.5. Broadly, there are five types—gas, solid-state, fibre, liquid and semi-conductor lasers.

Continuous wave lasers produce an output of relatively consistent power through delivery of uninterrupted beam of light. The wavelength of emission is dependent on the lasing medium. For example, CO_2 molecules *lase* at 10.6 μm while neodymium-based lasers produce stimulated emissions at 1047 and 1064 nm (figure 2.5). In cases where the lasing medium is an organic dye (liquid lasers), they are capable of generating a much wider range of wavelengths. This characteristic makes these types of lasers ideally suited to be tunable lasers, wherein the wavelength can be controlled during operation.

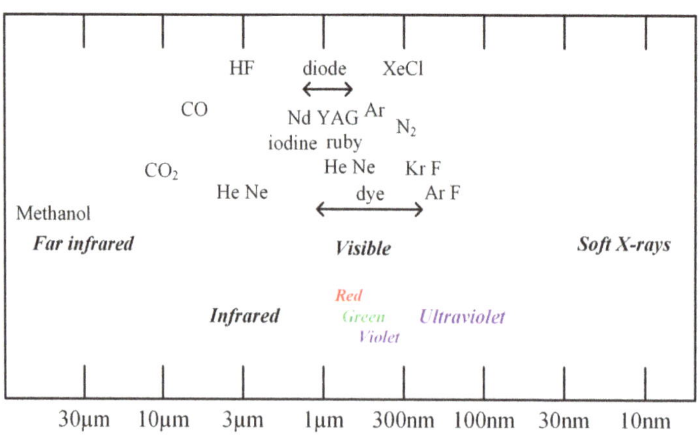

Figure 2.5. The electromagnetic spectrum-focussed illustration of the types of lasers, approximating their operating wavelengths.

Pulsed lasers[10] are a result of sustenance of laser action within the lasing medium, to a certain threshold. The result of this built up energy is a short pulse with high energy, sometimes in the order of picoseconds. Pulsed lasers, due to their time-dependent frequency of emission, find application in temporally resolved scientific experiments and in manufacturing due to the high powers these lasers can deliver (in the order of kilowatts to megawatts). Operating a pulsed laser is significantly different from the usage of a continuous wave laser. With regard to construction, once the aforementioned threshold is achieved, a 'gate' placed at the transmissive mirror is opened to generate the output from the laser cavity. Conceptually, the arrangement can be visualised as a setup similar to a continuous wave laser (figure 2.1), only adding a gate at the transmissive mirror to allow for build up of energy. This operation is known as *Q*-switching. Lasers that produce pulses in the range of 5 fs–100 ps (1 fs = 10^{15} s) are referred to as ultrafast lasers. It is considered to be the fastest man-made source to produce events, as this is faster than formation and dissociation of chemical bonds. For a detailed description of the various types of lasers, we direct the reader to the textbook, *Laser Physics* (Authors: Simon Hooker and Colin Webb. Oxford Master Series in Atomic, Optical and Laser Physics) [2].

2.3 Laser safety

A discussion of lasers and their applications cannot be complete, within its scope, without considering laser safety. As valuable as a laser has become to today's research and innovation ventures, they are and always will be dangerous to the human eye and, in some cases, skin. Without adequate protection and safety guidelines in place, the high power of the lasers can damage the vision of the operator and those directly in the path of the laser beam. A combination of personal laser safety equipment, laser warning signs and interlocks must be implemented for the safety of the researchers and others in the vicinity. Laser safety must be followed according to the laser system and power. Although this brief note introduces the user to the laser classes and their safety measures, it **must not** be considered a guide to laser safety to any extent. Laser safety requires dedicated attention and due diligence.

Laser systems are also classified on their output powers. Class 1 and 1M (<1 mW) lasers generate an output that is not harmful to the human eye and usually utilised in devices where the laser is inside. These lasers are relatively safe, even for long term intentional viewing, while Class 1M lasers could be harmful if viewed through focussing or magnifying optical instruments. Class 2 and 2M are lasers that emit visible light with a power less than 1 mW. A cautionary note of not staring into the beam for long durations is applicable, but the lasers are safe when unintentional exposure occurs, i.e. less than 1/4 second. Similar conditions apply for the 2M lasers, but as in the case of 1M lasers, optical instruments could enhance the laser beam and harm the eye.

[10] Commonly, a pulsed laser has emission pulse lasting about 0.5–500 ns, with much higher pulse rate constantly being achieved through latest innovations.

The power emitted by Class 3R lasers and above requires careful consideration of laser usage and safety. When working with 3R lasers, the power emitted in the visible wavelengths is less than 5 mW. Direct exposure to laser light must be avoided and unintentional exposure is relatively safe. It is good practice to ensure that there are signs on the door to indicate laser usage, but this does not need to be enforced.

Class 3B and 4 lasers emit less than, and greater than 500 mW, respectively. Warning signs outside experiment/usage room are compulsory and additional requirements like safety interlocks on doors must be enforced. These lasers are extremely harmful to the human eye, and this hazard is applicable not only to direct exposure but also reflected beam exposure. The lasers belonging to both these classes are capable of inducing thermal responses from the skin. While 3B lasers heat up the skin over long exposures, class 4 lasers are known to instantly burn the skin upon direct exposure.

2.4 Additional resources

A topic as diverse as lasers or formally known as *quantum electronics* cannot be justifiably discussed in a single text. Research and literature has evolved to include all forms of information and applications—governing physics, fundamental research, construction and materials, domain-specific applications, and many others. Along with the basic literature required to understand this phenomenal technology, ongoing research is also generating many volumes of information. At this stage, to the curious mind, we would like to share additional resources. Although most of the content here would be similar to those of the references, we would like provide all this information in a single location within the text.

- *Essentials of Lasers* (Author: L Allen) is a collection of texts describing lasers and their physics in the first half, while including the pivotal papers in the domain of quantum electronics in the second half. The reprints include the popular papers of T H Maiman (1960) for reference.
- *Fundamentals of Photonics* (Editor: C Roychoudhuri) is a freely available resource from SPIE (International Society for Optics and Photonics) made available to forward the knowledge base in the field of photonics. A chapter on lasers (Module 1.5) is available, authored by W T Silfvast.
- *Laser Physics* (Authors: M Sargent III, M O Scully and W E Lamb) is a textbook published by Addison-Wesley Publishing Company, Inc. focusing on the theoretical understanding of laser physics. The textbook is extensively inclined towards presenting a comprehensive understanding of the fundamentals of laser physics.
- *Lasers: Physics, Systems and Techniques* (Editors: W J Firth and R G Harrison) is a collection of talks presented at the Scottish Universities Summer School in Physics. Including dedicated introductory sections for lasers and their understanding, the articles proceed to more advanced concepts systematically.
- *Optics and Lasers: including Fibers and Optical Waveguides* (Author: M Young) is an introductory text published by Springer. The content of this book encompasses the basic concepts of light, its interactions with matter and delves into lasers as a part of an ongoing discussion in the field of optics.

It claims to be sufficiently populated to be a part of both an introductory course or a basis for an advanced academic program.

- *Introduction to Lasers and their Applications* (Authors: D C O'Shea, W R Callen and W T Rhodes) is a textbook designed not only for students, but also laboratory technicians and researchers in the domain of laser physics. It presents a holistic overview of the literature and concepts, sufficiently detailed for the novices and experienced alike.
- *The Laser Guidebook* (Author: J Hecht; Editors: R E Fischer and W J Smith) was presented as a part of the Optical and Electro-Optical Engineering Series by McGraw-Hill, Inc. The guidebook discusses the concepts of lasers and the history of their development, before proceeding to discussing each type of laser individually. The book was found be an asset during the formulation of the current literature in addition to the following textbook.
- *Lasers* (Authors: P W Milloni and J H Eberly) is a textbook cleverly divided into two sections. The initial chapters provide an overview of the field, its innate physics and applications. The second half of the book delves into the theoretical and practical aspects in detail.
- *Laser Physics* (Authors: S Hooker and C Webb) discusses laser physics, the interactions resulting in lasing action and the different types of lasers.
- *Quantum Photonics* (Author: T P Pearsall) discusses the fundamentals of electrons and photons from a quantum physics perspective. The book contains well described sections of lasers and lasing action, and entanglement and non-linear optics, among others.

References

[1] Pearsall T P 2020 Lasers *Quantum Photonics* (Cham: Springer) pp 193–226

[2] Hooker S and Webb C 2010 Introduction *Laser Physics* (Oxford Master Series in Physics) (Oxford: Oxford University Press) ch 1

[3] Maiman T H 1960 Stimulated optical radiation in ruby *Nature* **187** 493–4

[4] Maiman T H 1960 Optical and microwave-optical experiments in ruby *Phys. Rev. Lett.* **4** 564–6

[5] Allen L 1969 Crystalline solid lasers *Essentials of Lasers (The Commonwealth and Int. Library: Selected Readings in Physics)* ed D Ter Herr (Oxford: Pergamon) ch 3 pp 19–27

[6] Hecht J 1992 Introduction *The Laser Guidebook (Optical and Electro-Optical Engineering Series)* ed R E Fischer and W J Smith 2nd edn (New York: McGraw-Hill) ch 1 pp 1–12

[7] Allen L 1969 Introduction to the laser *Essentials of Lasers (The Commonwealth and Int. Library: Selected Readings in Physics)* ed D Ter Herr (Oxford: Pergamon) ch 1 pp 3–7

[8] Milonni P W and Eberly J H 1988 *Lasers* (Wiley Series in Pure and Applied Optics) (New York: Wiley) ch 1

[9] O'Shea D C, Callen W R and Rhodes W T 1978 Laser light *An Introduction to Lasers and Their Applications* (Addison-Wesley Series in Physics) (Reading, MA: Addison-Wesley) ch 2 pp 5–50

[10] Hecht J 1992 Laser theory and principles *The Laser Guidebook* (Optical and Electro-Optical Engineering Series) ed R E Fischer and W J Smith 2nd edn (New York: McGraw-Hill) ch 3 pp 23–42

IOP Publishing

Laser Beam Profiling: Cost-Effective Solutions

Akhil Kallepalli and David B James

Chapter 3

Silicon-based imaging systems

Abstract

The human eye is the most primitive of imaging systems. Most imaging systems today have an analogous construction to that of an eye. When you deconstruct both, there is an aperture to allow and control the amount of light entering the system, a lens to focus the light and a photosensitive detection surface.

Silicon has emerged as an indispensable material in electronics and photonics, and is vital in manufacturing semiconductor devices. The spectral sensitivity of silicon is comparable to the human eye, and therefore is perfectly suited for manufacturing detectors that work in the visible range of the electromagnetic spectrum.

In this chapter, we will discuss the human eye and its working, before proceeding to discuss charge-coupled device (CCD) and complementary metal–oxide–semiconductor (CMOS) sensors. Finally, the characteristics of silicon-based detectors are discussed. It is important to note that semiconductor-based devices, and indeed each of these devices, are a field of study on their own.

When light is incident on a surface, four interactions occur to varying degrees: the energy is reflected, scattered, absorbed and/or transmitted. This interaction continues to hold good for materials used for detection of light energy. This incidence of radiation acts as a stimulus for a response. In the case of the human eye, this response is a signal to the brain. This signal is analogous to a video stream from the eye to the brain. Similar interactions with materials can be converted to signals that can be measured, and representations of the scene [1]. In this chapter, we will briefly discuss the human vision and perception, before moving ahead discussing silicon sensors, i.e. CCD and CMOS devices.

3.1 Vision

Human vision is an optical system projecting an image through a lens and several layers of neurons onto a complex system of distributed photoreceptors. While this

doi:10.1088/978-0-7503-3835-6ch3

description, in its basic form, is correct, let's look at it in a bit more detail. The human eye receives light through the combination of the cornea, iris and the lens (figure 3.1). The cornea's primary function is to bend light and ensuring the majority of incident light works its way through the lens and onto the retina, eventually. The iris, working as aperture, regulates the amount of light entering the eye. It contracts or dilates the pupil depending on the amount of light falling on the retina, as a response to controlling brightness. Subsequently, the lens focuses the light that has entered the eye through the vitreous humour and onto the fovea (figure 3.1). The fovea is a small cavity on the retina beyond which passive cone photoreceptors reside. Currently, the most widely accepted vision theory suggests that humans have a tri-chromatic system that responds to incident light and can separate the three primary colours. This is made possible due to specific, different types of band pass cone-shaped absorbing photoreceptors in the retina [2, 3]. The 'fragments' gathered by the retina are combined by the brain, through its ability to combine separate entities into one, logical image.

The human retina is estimated to contain 110–125 million rods and 6.3–6.8 million cones. These two types of photoreceptors in the eye serve different purposes, i.e., scotopic (nocturnal) and photopic (diurnal) abilities of sight. The cones are

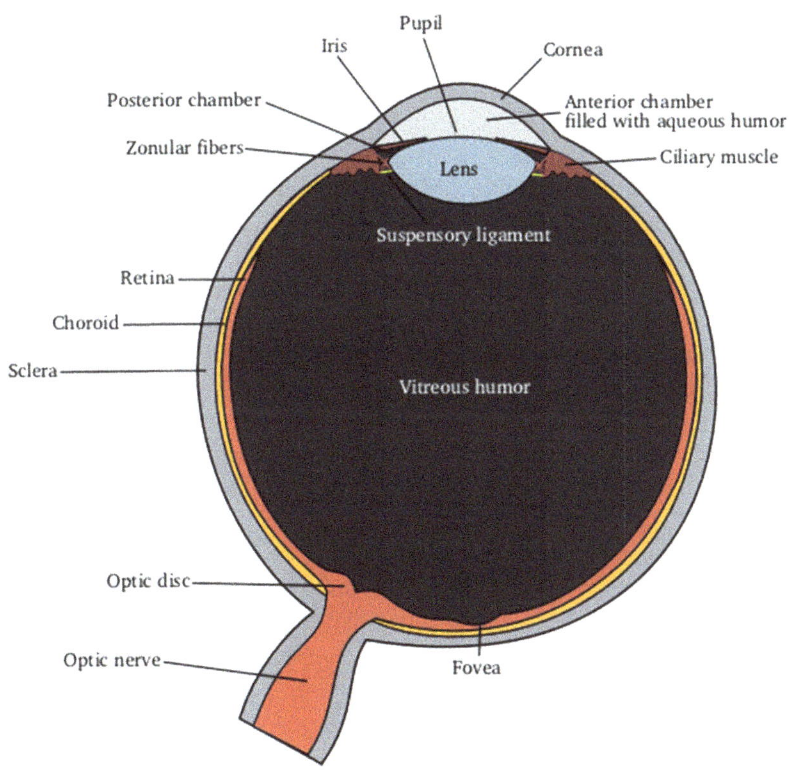

Figure 3.1. The schematic of the human eye. The light is focused by the lens, through the vitreous humor onto the retina. This 'Schematic_diagram_of_the_human_eye' image has been obtained by the author(s) from the Wikimedia website where it was made available by Delta G under a CC BY-SA 3.0 https://creativecommons.org/licenses/by-sa/3.0/deed.en.

highly concentrated in the central region of the fovea and are responsible for capturing a high resolution image for the brain. Rods, however, are present around the cones while eventually dropping in distribution towards the periphery. Due to this distribution, they are responsible for peripheral vision and result in poorly resolved images, relatively. The rods and cones in the eye serve different roles and function differently. These are:

1. Rods function in dim light while cones work in daylight and bright conditions.
2. Rods are highly sensitive to light while cones are not. This higher sensitivity is key as it allows the eye to focus on any objects that are detected as being in the peripheral vision, if needed.
3. Cones are capable of resolving an image far better than rods.
4. The temporal resolution of cones is also better than that of the rods.
5. The maximum sensitivity of rods is in the blue-green region of the electromagnetic spectrum, while cones' range from yellow to green.
6. The tri-chromatic nature of human vision, i.e., the ability to see colour is possible due to cones as rods are colour-insensitive.

The combinations of responses of the optical stimuli detected by the eye are sent to the brain for interpretation. This can be described as a video stream where the aperture (pupil) defines the amount of light entering the eye and the combination of rods and cones detect a highly resolved object from a poorly resolved background.

Our visual abilities today are a result of various optical phenomena occurring in front of, and within the eye before light reaches the retina. Absorption, reflection and scattering mechanisms are constantly in operation and the 'visible' region of the spectrum is detected when incident on the retina. This region begins at approximately 420 nm and ends at 700 nm, spanning what we recognise as blue to red colours, respectively. Using known quantities such as the field of view achieved by the eyes, one can quantify the resolution of the human eye. The acuity of the cones and rods, combined with the 120° field of view results in a 340×10^6 pixels for each eye. Having two eyes for collecting incoming information that is processed in real time by the brain improves the resolution of the scene being perceived in every single snapshot. To truly understand how incredible the human eye is, consider that (1) the eye is capable of adjusting to changes in brightness and reach its full resolution potential almost instantly in comparison to sensors taking seconds, even minutes and (2) the best of imaging technology available in current mobile phones is only 108 MP; a third of what an each eye is capable of.

3.2 Detectors

Any device capable of responding to a stimuli and converting the information into a measurable quantity can be considered a detector. In this text, we will discuss optical detectors which respond to light and convert it into electrical signals. Further, we will focus on detectors using semiconductor materials (also known as solid-state detectors). Such detectors are usually composed of two primary components:

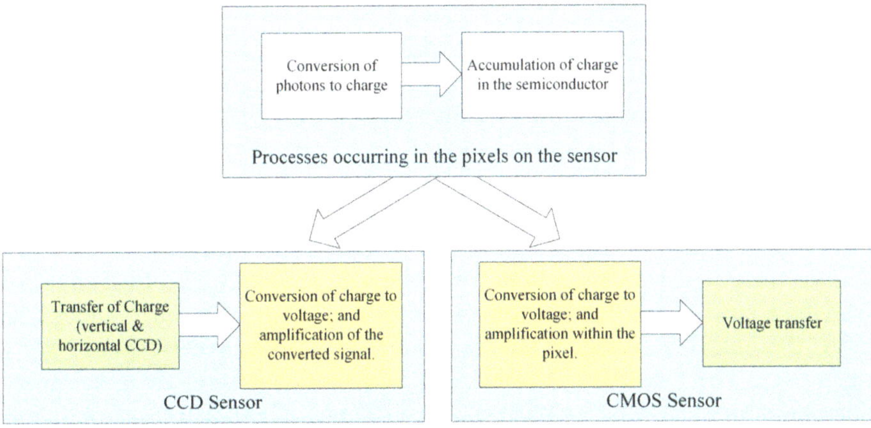

Figure 3.2. Workflow of CCD and CMOS imaging sensors illustrating the similarity of the processes of image acquisition, while showing their differing sequences.

1. A collection of photosensitive, closely placed detector elements that convert light to electrical signals.
2. An electrical readout circuit that relays this signal to output amplifiers.

Most optical detectors respond to photons that have an energy greater than the bandgap of the detector material (figure 3.2). In the case of silicon, visible wavelengths and a portion of the near-infrared region of the spectrum are capable of inducing this photoelectric effect, thus resulting in a proportional electrical signal. This electrical signal is proportional to the rate of absorption of the photons incident on the detector surface.

In addition to the detector materials, a variety of detector constructions are also available (figure 3.2). These are mainly divided into three structures [4]:

1. Photoconductive detectors; the absorbed photons in these detectors cause a change in conductivity of the detector. A constant bias is placed across a detector and the modulation in the current is proportional to the change in incident light.
2. Photovoltaic detectors; the absorbed photons in these detectors induce a change in voltage at a potential barrier that is either produced by a p–n junction or electric fields with a metal–insulator–silicon (MIS) structure.
3. Photoemissive detectors; such as silicide Schottky barrier devices wherein the photons absorbed by the silicide electrode induce excited carriers that are emitted over the Schottky barrier into the semiconductor. The charge generated at the electrode is transferred to a readout structure.

The differences continue as the readout electronics can also be designed in many ways, each with their own advantages and disadvantages.

1. Charge-coupled devices (CCD) readouts.
2. MOSFET switch readouts such as CMOS circuits.

3. Charge injection device (CID) readouts.
4. Charge imaging matrix (CIM) readouts.

While combinations of readout circuits and detector types have been fabricated, below 1100 nm, the most common and easily available detectors are silicon based, operating as photovoltaic detectors. Two of these, CCD and CMOS sensors are discussed in this chapter and are relevant to laser beam profiling in the visible and near-infrared wavelengths (figure 3.2).

Photons absorbed by silicon detectors induce a photoelectric effect that creates electron–hole pairs. When the photon, of energy hc/λ, is incident on a semiconductor material, the cutoff wavelength (in μm) that can be detected is defined by 1.24/bandgap (eV). The bandgap of silicon is 1.1 eV and therefore, these detectors are sensitive to photons of wavelength less than 1100 nm. These detectors are limited at the other end of the spectrum at blue/UV wavelengths because the absorption length is too short for the photon to reach the silicon beyond the surface layers and induce the necessary photoelectric effect.

3.3 Charge-coupled devices (CCD) detectors

In silicon detectors, there are two basic types of readout types: charged-coupled devices (CCDs) or metal–oxide–silicon (MOS) X–Y addressed switch arrays. In a CCD, the charge generated by the incident light is moved across the chip towards an amplifier by applying a clocking sequence to the electrodes [4].

To implement the concept of free electrons' generation and the subsequent charge transfer, CCDs emerged from the research into 'silicon-based electrical equivalent of the magnetic bubble memory' [5]. The basic schematic of the CCD (figure 3.3)

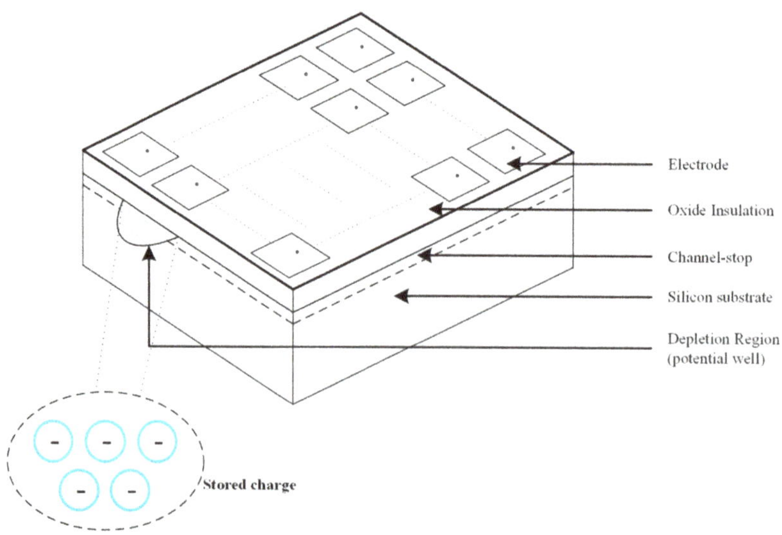

Figure 3.3. Structure of the charge-coupled device (CCD).

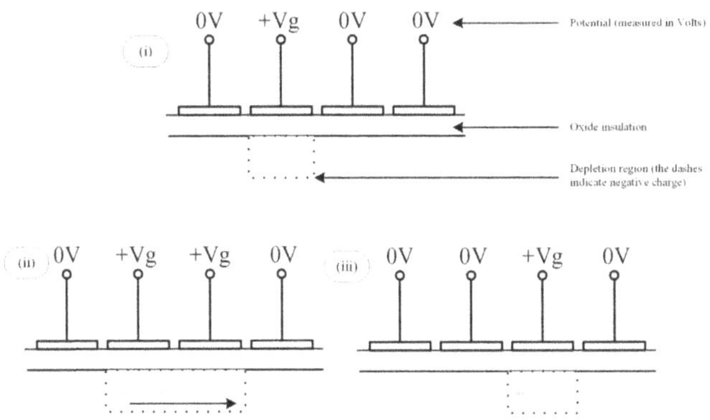

Figure 3.4. Charge transfer along the electrode array.

illustrates an arrangement of a series of electrodes, closely spaced, separated from the underlying semiconductor substrate by an oxide layer for insulation. When compared to the previously used camera tubes for imaging, CCDs offer portability, low power consumption and compactness for imaging systems. Silicon, used in both CCD and CMOS systems, is very sensitive to incident light and, in some special cases, is capable of detection of almost all the photons incident on it. The applicability of the silicon sensors can also be enhanced to sense ultraviolet and soft x-rays through specialised fabrication and optimisation.

The basic structure of CCDs involve an orthogonal arrangement of electrodes and charge transfer channels. Charge transfer (figure 3.4 illustrates a one-dimensional charge transfer process) occurs when incident light charges the electrodes on the sensor and the electrodes are pulsed or clocked to transfer the image charge pattern to the receiving array, one line at a time. The lowest row or column on the sensor, closest to the serial readout register, is transferred first. As the register is clocked separately and faster than the pixels, the information in the sensor is readout and transferred to the memory quicker than the exposure measurement at the pixels and the next line of image data can be transferred. With the design, there is an inherent loss of information due to smearing of the images during readout. The pixels were exposed to the incident photons, while being read out to the register, thereby defeating the purpose of image acquisition. The inherent electronic shutter of CCD sensors easily circumvents this problem, wherein the image was read to the memory in darkness (i.e. the shutter is closed during read out). Many advances have occurred since the fabrication of the first CCD sensor, enhancing its charge-transfer efficiency, reducing the readout noise and extending its spectral range and efficiency. Scientific CCDs have also emerged, with specific applications to research that demands higher sensitivity and the detection of almost all the incident light is a bare necessity.

The charge transfer process has evolved with changes in the readout architecture. Four types of readout architecture now exist, each with their advantages and challenges (figure 3.5). These are (1) full frame, (2) frame transfer, (3) interline transfer, and (4) frame interline transfer CCDs. Of these, full frame CCDs are the

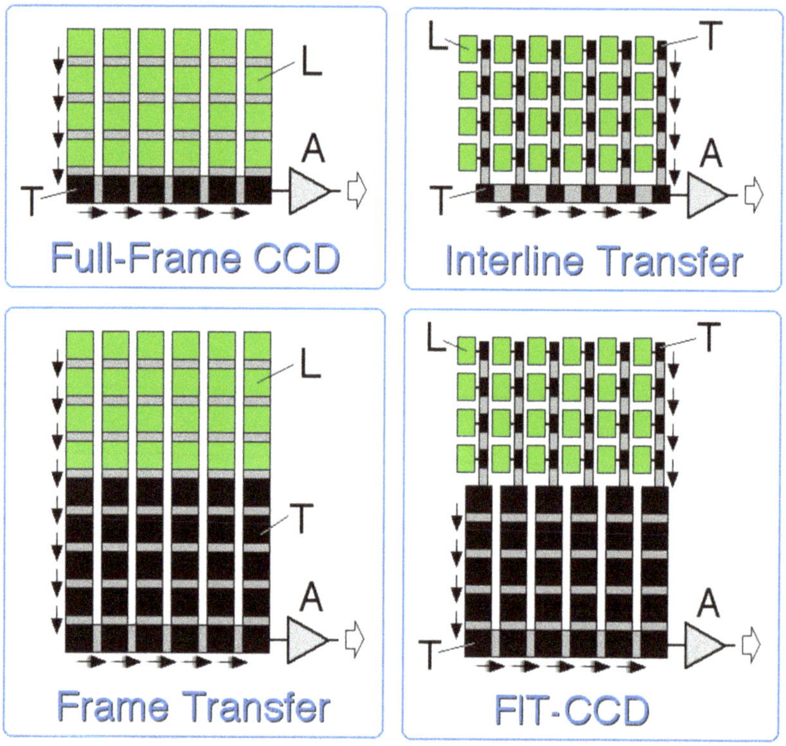

Figure 3.5. The readout architecture of CCDs define the characteristics of the imaging device and the challenges/advantages it presents. The FIT and interline transfer CCD devices are complex and have higher manufacturing costs and lower sensitivity due to smaller photosensitive area. Simultaneously, frame transfer and full frame devices use a larger photosensitive area but tend to require amplifiers to reduce readout time. The choice of architecture is purpose-defined. This 'Types of CCD image sensors, schematic' image has been obtained by the authors from the Wikimedia website where it was made available by Michael Schmid under a CC BY-SA 1.0 https://creativecommons.org/licenses/sa/1.0/.

simplest to make and operate. Images are optical projected on the parallel array and the rows are sequentially transferred to the output amplifier. The simplicity of this design yields the highest possible resolution for a given chip size, but with the downside of slower frame rates as the each frame is only acquired after the previous one has been completely transferred to the amplifier.

Frame transfer CCDs are similar in architecture to full frame CCDs, but with a reduced photosensitive area due to the presence of both image array and storage arrays on the chip. While faster frame rates are achieved, they do suffer from a reduced resolution (and higher costs) as the photosensitive area is reduced by half to accommodate for the storage array.

Interline transfer CCDs addresses the shortcomings of the frame transfer by dedicating the readout transfer of each photosensitive surface. After integrating a scene, the signals of the pixels are collected simultaneously, allowing the detector to be ready for the next frame. The disadvantages of previous architectures such as image smear and slower frame rates are not found due to the continuous operation of the sensor.

Finally, frame interline transfer combines both interline transfer and frame transfer architectures and achieves the advantages without the problems such as smear and slower frame rates. However, the primary challenge in this case is fabrication costs and lower sensitivity for a given sensor size. Lower sensitivity occurs due to comparatively less photosensitive area at each pixel, with detector space being occupied by the readout circuit. Additionally, quantisation or sampling errors are greater due to the reduced aperture and some architectures suffer from image lag as a consequence of charge transfer from the photodiode to the CCD. A constant trade-off must be considered between frame rates, resolutions and overall efficiency of the detector based on the architecture of choice.

3.4 Complementary metal–oxide–semiconductor (CMOS) detectors

Active-pixel sensors are a class of imaging sensors that contain an array of pixels, each with its dedicated photosensitive area and signal amplifiers. Until the evolution of the CMOS (a type of active-pixel sensors) sensors, the imaging value of the CCD sensors was unmatched. The emergence of the CMOS sensors radically changed the resolutions achieved and the affordability of the sensors. The major advantage over the CCDs was that multiple transistors could be integrated on a single silicon chip-based circuit. The advantage extended to being able to integrate a large array of pixels to their own photodiode and readout transistors, alongside the additional electronics required to operate the sensor, read out the captured intensities into buffers and even digitise it for processing and/or storage. This allows CMOS sensors to perform digitisation at every pixel itself. CMOS sensors also operate at voltage lower than CCDs, therefore requiring less power for successful operation.

In comparison to the CCDs, modern CMOS sensors have a very different structure. Each pixel consists of a light sensitive region, a signal amplifier and a pixel-select switch. The photons generate charge that is stored in the semiconductor substrate, like in CCDs. As the charge generated by the photons is very small for transfer along the microwires, the conversion to voltage and amplification is performed at the pixel itself. In contrast with the line-based charge transfer in CCDs, each microwire carries the signal from individual pixels through the pixel-select switch. This characteristic allows *windowing* wherein a part of the sensor can be read out instead of the whole sensor due to the ability of being able to read individual pixels off the sensor. Following a precise sequence, pixel-select switches and column-select switches are turned on from top to bottom, and left to right, respectively. As this sequence is repeated for every pixel in the sensor, the image is acquired and the voltages are transferred to the signal processor of the camera [6].

CMOS sensors are considered to be more robust in their utilities due to their compactness, lower mass and lower power requirements. They could also be more radiation tolerant, which is a requirement for applications like imaging in space. Recent advancements have also made the initially poor light sensitivity of the CMOS sensors a property of the past. The CMOS sensors also present the

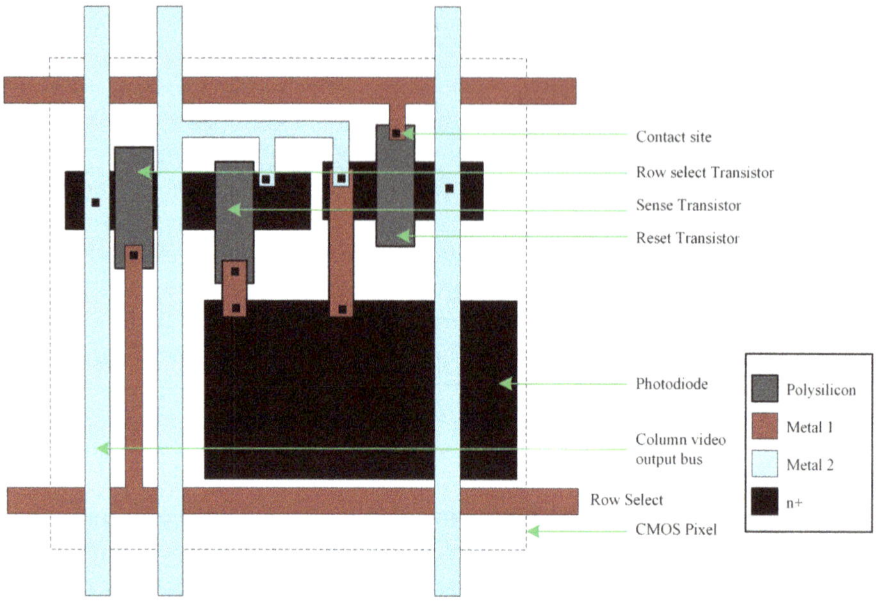

Figure 3.6. Complementary metal-oxide semiconductor (CMOS) pixel and three transistors layout.

advantage of windowing. The CMOS sensors today also offer significantly high frame rates in comparison to CCDs, when considering the cost-versus-performance ratio. This was achieved by the improved response to light and global shutter features. Image artefacts like *blooming* and *smearing*[1] do not occur when using CMOS sensors for imaging. On the downside, CMOS sensors do not have a large *fill factor*. The fill factor refers to the fraction of the pixel sensitive to light [5]. As seen in figure 3.6, the area occupied by the photodiode is approximately 19%, achieving a fill factor of 30%. Losses in signal strength and conversion to information occur due to reflections from the metallic (usually aluminium) regions of the sensor. These losses are also being mitigated by employing a back-illuminated sensor approach.

The CMOS readout architecture has advantages such as reducing cost, simplifying the clock requirements and allows to have additional on-chip circuitry such as digitisation, edge detection, thresholding, motion detection and spatial filtering [4]. These devices are ideally suited to low-cost imaging and machine vision applications, and their performance can, in some conditions, rival that of scientific CCDs.

3.5 Detector characteristics

When deciding an imaging sensor for an application, the conditions of usage and requirement from the data plays a key role. Although traditional CCDs had better sensitivities in low light conditions, CMOS technologies have fast evolved to

[1] Blooming occurs when saturated pixels in a CCD leak light into the surrounding pixels, while smearing occurs when charge transported through the CCD sensor in bright light conditions.

comparable and, in some cases, better performance. Certain parameters, discussed briefly, must be considered when choosing an imaging sensor.

3.5.1 Sensitivity

Sensitivity of an imaging system is defined using two quantities—quantum efficiency (*QE*) and read noise (*RN*). Quantum efficiency is a measure of the conversion efficiency of the sensor from photons to electric charge (electrons). A *QE* of 100% represents a conversion efficiency[2] of one electron per incident photon (on average). The read noise is the noise generated by the sensor in the dark and at zero integration time. In addition to this, the sensor also experiences *dark current* which is the signal generated from the sensor when no light is allowed to fall on the sensor surface. Therefore, to quantify the sensitivity of the sensor *at a given wavelength*, the measure of *QE*/*RN* is used. The higher this value is, the higher is the sensitivity of the sensor at a given wavelength. The sensitivity (or response) of silicon as a function of wavelength (figure 3.7) illustrates its value and utility for designing imaging sensors in the visible wavelength range.

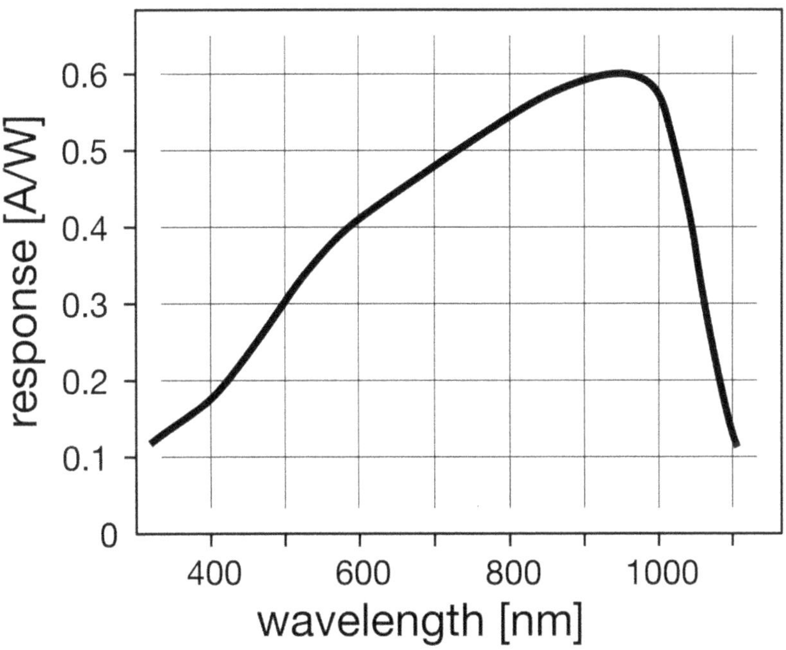

Figure 3.7. Silicon's spectral sensitivity as a function of wavelength shows its utility as a material for manufacturing photosensitive detectors in the visible (and parts of the near-infrared) wavelengths. This 'Response of a typical photo diode vs wavelength of the incident light' image has been obtained by the authors from the Wikimedia website where it was made available by KaiMartin under a CC BY-SA 3.0 https://creativecommons.org/licenses/by-sa/3.0/deed.en.

[2] A quantum efficiency of 60%–70% in the green wavelengths is most common in imaging systems today. However, optimised sensors for specific applications are available.

3.5.2 Spectral response and detector responsivity

Spectral response of a sensor refers to the spectral region or electromagnetic wavelengths it can 'see'. The response of the detector when light is incident on it is known as its responsivity. This quantity defines the efficiency of the device, quantifying it as a ratio of output current to input optical power. The amount of time the sensor takes to respond to variations in input light intensity is another parameter defining the sensor's applicability; known as *response time*.

3.5.3 Noise

In terms of producing noise during acquisition, CCD sensors have long had the advantage of better performance. However, using the technology of buried photodiodes, CMOS sensors reduced the difference between the two sensors, enabling both platforms to produce clear and low-noise images. The noise generation of the sensor also plays a role in the performance of the sensor in low light conditions. A common method to remove noise from the images is to acquire a dark frame, i.e. an image with the shutter/aperture of the system closed (figure 3.8). This dark frame, when subtracted from any image acquisition, removes the sensor's noise from the image.

Figure 3.8. Dark frames as CCD acquisitions that are taken with the shutter closed, i.e., no light is entering the system. This records the dark current on the CCD chip and illustrates 'hot pixels' that appear scattered over the frame. Acquiring this dark frame sets a baseline for subtraction from any images taken with the shutter open, thereby removing the sensor noise. This 'CCD_Dark_Frame' image has been obtained by the authors from the Wikimedia website where it was made available by H Raab (User:Vesta), Johannes-Kepler-Observatory, Linz, under a CC BY-SA 3.0 https://creativecommons.org/licenses/by-sa/3.0/deed.en.

Figure 3.9. Smear is an undesired signal that appears as a bright vertical stripe, emanating from the brightest part of the image. Vertical smears in CCDs occur due to the incident light during the image shifting from the sensor to the storage area (frame transfer CCDs) or scattered photons tunnelling into the darkened shift register (interline transfer CCDs) instead of activating the photosensitive area. Smear in CMOS sensors is due to the same reason as interline transfer CCDs. This file is licensed under the Creative Commons Attribution-Share Alike 4.0 International license https://creativecommons.org/licenses/by-sa/4.0/deed.en.

3.5.4 Vertical smear

CCDs are prone to vertical smears, wherein the light leaks into the vertical shift registers. Stray light, converted into signal, presents itself as bright streaks above and below (figure 3.9). In comparison, CMOS sensors are immune to vertical smear as they use signal wires to transfer the voltage. When light is incident on these, it does not create any erroneous signal. CCDs have had technological advancements that have ensured that smear artefacts have been reduced by using micro lenses and reducing the thickness of the insulating layer, preventing stray light.

3.5.5 Resolution

As a general rule of thumb, the size of the pixels in a sensor must be twice as large as the wavelength they are intended to sense. Therefore, limited by the size of the sensor and the pixel (expressed in micrometers (μm)), the resolution of a sensor is defined in terms of height (vertical pixels or columns) and width (horizontal pixels or rows) of the sensor surface, mentioning the size of the pixel as well. In some cases, they are referred to as *sensor format* and given in inches. This measure is the diagonal length of the sensor.

While this resolution defines the sensor, it is important to note that other resolution references are also prevalent. For instance, a satellite image taken at a higher altitude is considered to be poorly resolved in comparison to that from a lower altitude (using the same sensor system). Images taken from a microscope with a higher magnification objective are considered to have a better resolution than one with a lower magnification. In each case, the definition of resolution as a parameter changes based on the field of research or topic of discussion.

3.5.6 Exposure mode and frame rate

Light is incident on the sensor continuously, until a change of intensity occurs or the source of light ceases to emit radiation. However, a constant stream of photons makes it impossible to account for time needed to read the accumulated charge in the pixels and transfer it, through voltage, to the imaging system to generate an image. To accommodate for this time, mechanical or electronic shutters are employed. The function of these shutters is to block the incoming photons to allow the pixel to be read, prior to the next acquisition. There are two types of shutters—rolling and global shutter. A rolling shutter collects images by rows, thereby collecting information from top to bottom and remaining 'dark' until the next exposure. The analogy could be made to that of a wave motion. The only case where this method of acquisition creates artefacts is when the sensor is attempting to capture a fast moving object. As the pixel rows are initiated for acquisition consecutively, the object may move or shift with respect to the sensor. This causes a 'tilt' artefact in the capture. In comparison, global shutters expose the entirety of the sensor to the incoming light at a single instant, making it the best representation of capturing objects in motion. However, the global shutters enhance the image acquisition at the cost of reduced frame rates in comparison to the high frame rates of rolling shutter sensors; the frame rate is the number of complete acquisitions the sensor is capable of capturing per second.

3.6 Additional resources

A topic as diverse as lasers or formally known as *quantum electronics* cannot be justifiably discussed in a single text. Research and literature has evolved to include all forms of information and applications—governing physics, fundamental research, construction and materials, domain-specific applications, and many others. Along with the basic literature required to understand this phenomenal technology, ongoing research is also generating many volumes of information. At this stage, to the curious mind, we would like to share additional resources. Although most of the content here would be similar to those of the references, we would like provide all this information in a single location within the text.

- *High Performance Silicon Imaging* (Editor: D Durini) is a highly informative text looking at the silicon sensors technologies with detailed chapters about the development of CCD and CMOS sensors. The text also highlights the recent applications where these technologies have been applied.

- *A Guide to the Use and Calibration of Detector Array Equipment* (Authors: G R Hopkinson, T M Goodman and S R Prince) shares detailed information about silicon technologies and their applications. Further, their calibration is the primary focus of this book. These techniques are important when using these technologies for applications.

References

[1] Durini D and Arutinov D 2014 Fundamental principles of photosensing *High Performance Silicon Imaging* ed D Durini (Cambridge: Woodhead Publishing) ch 1 pp 3–24

[2] Curcio C A, Sloan K R, Kalina R E and Hendrickson A E 1990 Human photoreceptor topography *J. Compar. Neurol.* **292** 497–523

[3] Rosenthal E, Solomon R J and Johnson C 2004 Waves vs. photons: A newview of how we detect light and color *Technical report* Creative Technology

[4] Hopkinson G R, Goodman T M and Prince S R 2004 Detector arrays *A Guide to the Use and Calibration of Detector Array Equipment* (Bellingham, WA: SPIE Press) pp 1–55

[5] Waltham N 2013 *CCD and CMOS Sensors* (New York: Springer) pp 423–42

[6] Sony Corporation 2008 CCD and CMOS Image Sensor Technologies: A Comprehensive Guide for Professional Videographers. Published as a white paper/brochure accessed through an internet search on 28-09-2017

IOP Publishing

Laser Beam Profiling: Cost-Effective Solutions

Akhil Kallepalli and David B James

Chapter 4

Experimental techniques

Abstract

Having discussed, briefly, lasers, CCDs and CMOS sensors, this chapter is the experimental chapter that provides the reader with the technique and skills in the laboratory to profile a laser beam. Monitoring the laser, before and after, interaction with an object or medium could potentially provide valuable information about the object/medium based on the changes in the beam profile.

In this chapter, we will share the devices needed for profiling a laser beam and detail the methodology. Before beginning, the readers are advised to download the computer code and templates shared online, and download the video files provided with this book (available at https://iopscience.iop.org/book/978-0-7503-3835-6) that show the experimental procedure. These guides are key to profiling a laser beam.

The methodology detailed in this chapter can be integrated in any experiment where the beam profile needs to be monitored or recorded.

Laser beam profiling is a key step in any experimental set-up. Understanding the input or the output beam (or both) in many cases is necessary to estimate the interactions that light has with objects or media. To do this in a simple and safe manner, we present a laboratory experiment in this chapter. We will discuss the set-up of equipment, stages of the measurement and the associated code and results. Through the course of this analysis, there will be references to snippets of code from the program, stating their function and importance. This code can be downloaded[1], along with the relevant data formatting from the MATLAB® Community page [1]. A demonstration of this experiment is available in video files 1–5.

The equipment used for laser diagnostics are (figure 4.1):

1. Laser (Diode Module CPS520; Thorlabs, Inc.);
2. Laser power meter (ADM 1000; Lasermet) and display unit;

[1] Download Link (accessible through digital copy at https://uk.mathworks.com/matlabcentral/fileexchange/64402-laser–filters-and-cmos-imager-calibration). If the link is inaccessible, search for 'Laser, Filters and CMOS-Imager Calibration' on MATLAB File Exchange.

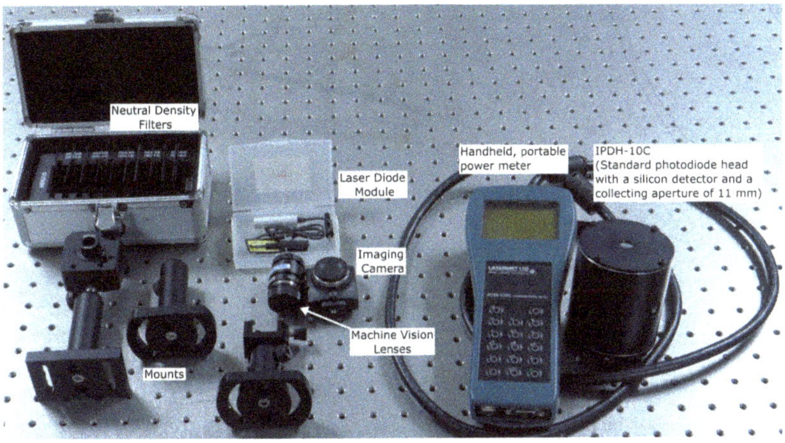

Figure 4.1. For profiling a beam, any CCD/CMOS-based imaging system can be used. To measure the power and the profile of the beam, you will need a laser, a power meter, neutral density filters, an imaging system and mounts.

3. Neutral density filters (NE2xxB series; where xx signifies the optical density of the filter.; Thorlabs, Inc.);
4. Imaging sensor (DCC1545M; USB 2.0 CMOS camera with a resolution of 1280×1024 pixels.; Thorlabs, Inc.).

The laser diagnostic workflow will involve a measurement of the laser power at a desired distance (figure 4.2(i)), measuring the laser power through individual neutral density filters with different optical densities (figure 4.2(ii)) and using the reduced power to plot the beam's profile using an imaging system (figure 4.2(iii)). The laser, in this context, is a Class 3R laser (laser safety described in section 2.3 should duly be observed) operating at or under 5 mW.

The laser power measurement is the first step of the calibration procedure. Most manufacturers measure the laser power prior to shipping to the customer. However, in operating conditions which are in most cases different from the calibration conditions, the laser power can deviate from the given measurement. Therefore, it is imperative to test and calibrate the laser power in the experimental conditions to gain a better understanding of the laser's output. A single CCD-based photodiode head with an integrating sphere collects the laser beam and measures its power. The integrating sphere successfully reduces the number of reflections, while assisting in easy alignment with the laser. The system, Lasermet ADM 1000, is highly accurate and is effective with a response time of 20 ms.

Once the laser power has been measured, the neutral density filters must be calibrated to understand how the operating conditions affect the transmission through the filters. Yet again, manufacturers are known to pre-calibrate the neutral density filters. However, when using a laser of a specific wavelength, the transmission through the filter is known to vary. Extensive use of the filters also induces damage on the filters, which can also impact the optical density of the filters. The neutral density filters are used consecutively, placed in the path of the beam between the laser and the

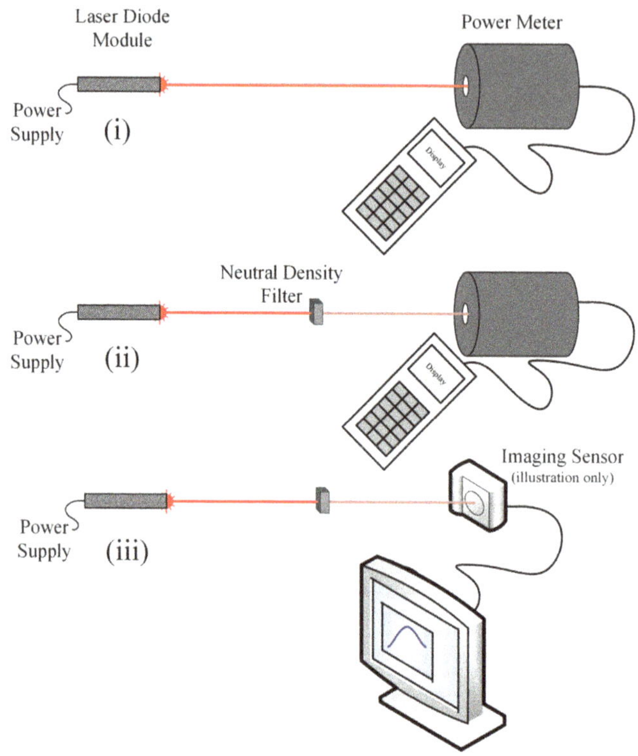

Figure 4.2. Application of the laser power meter, neutral density filters (to reduce the power) and the imaging system for the beam profile and intensity measurement.

power meter. The transmission of the filters can be measured in the laboratory conditions and quantified for further application when using the imaging system.

While operating at a power that is of a relatively low danger to the human eye or skin, the laser power is still large enough to cause thermal and/or physical damage to the imaging sensor. Therefore, the use of neutral density filters to reduce the intensity of light incident on the sensor is a key to profiling the beam without damaging the sensor. The beam profile is captured on the imaging sensor, processed using MATLAB to generate a three-dimensional surface plot to better understand the nature of the laser.

4.1 Experimental workflow

Lasers emit stimulation radiation with a very high degree of directionality. Therefore, for any experiment involving a laser system, alignment is the key. For this purpose, we use rails and mounts to set up the experiment. The laser calibration follows the workflow detailed. With every laser experiment, due consideration must be given to the reflective surfaces around the experimental arrangement. For this experiment, it is also imperative to work in a dark room. Realistically, it could be challenging to set up a 'true' dark room. However, using the power meter and the imaging system, at a later stage, the pre-existing conditions of the room can be quantified.

4.1.1 Laser power measurement

The laser diode module, manufactured by Thorlabs Inc. is used as the light source in this methodology. With a proper understanding of laser powers and neutral density filters, most lasers can fit the purpose for laser diagnostics using imaging systems. In this scenario, we are implementing the method using the Class 3R CPS520 diode module that is emitting stimulated radiation at 515 nm (λ_{CW}).

1. The laser diode module is mounted on the rail, connected to the power supply and switched on. Every laser system has a **'warm up'** time, which in this case is less than two minutes. It is good practice to allow the laser to reach its optimum performance prior to experimentation.
2. Once the laser is ready for use, the **power meter** is oriented in the horizontal direction and aligned to the beam, allowing the entire beam to enter the integrating sphere head. The power meter connected to the photodiode allows the input of the wavelength of inquiry. Using the calibration information specific to the laser module provided by the manufacturer, the power of the laser at five wavelengths above and below the central wavelengths was measured. Another important note at this stage would be to test the laser power meter at the wavelength of inquiry in the dark room, with the laser power switched off. This will provide the operator with an idea of the light in the room[2].
   (a) In this methodology (figure 4.2(i)), the Lasermet ADM 1000 power meter was used to measure the CPS520 (λ_{CW} = 515 nm[3]) laser diode module's power. As the central wavelength of the diode module is given, the power at wavelengths from 505 to 525 nm with increments of 1 nm is measured in order to visualise the power distribution as a function of wavelength.

The measured laser power in the range of the wavelength (505–525 nm) is fitted to a linear model using the ***fitlm()*** function in MATLAB (figure 4.3). The plot shows a relatively linear relationship between the power and the wavelengths.

4.1.2 Neutral density filters' calibration

Neutral density filters are available with varying optical densities, allowing different degrees of transmission. This is achieved by using materials/coatings that absorb light in the wavelengths of operation. Neutral density filters are usually calibrated at individual wavelengths, illustrating transmission at various wavelengths. The transmission percentage (T) of the filter is governed by its optical density (OD), measured using the relationship (equation (4.1))

[2] During the calibration procedure at the time of this writing, the room's power was measured to be below 1 μW, while the power of the laser was over 1 mW. Therefore, it could be disregarded in calculations.

[3] A calibration sheet is provided with product-specific calibration by Thorlabs Inc. with the laser diode module.

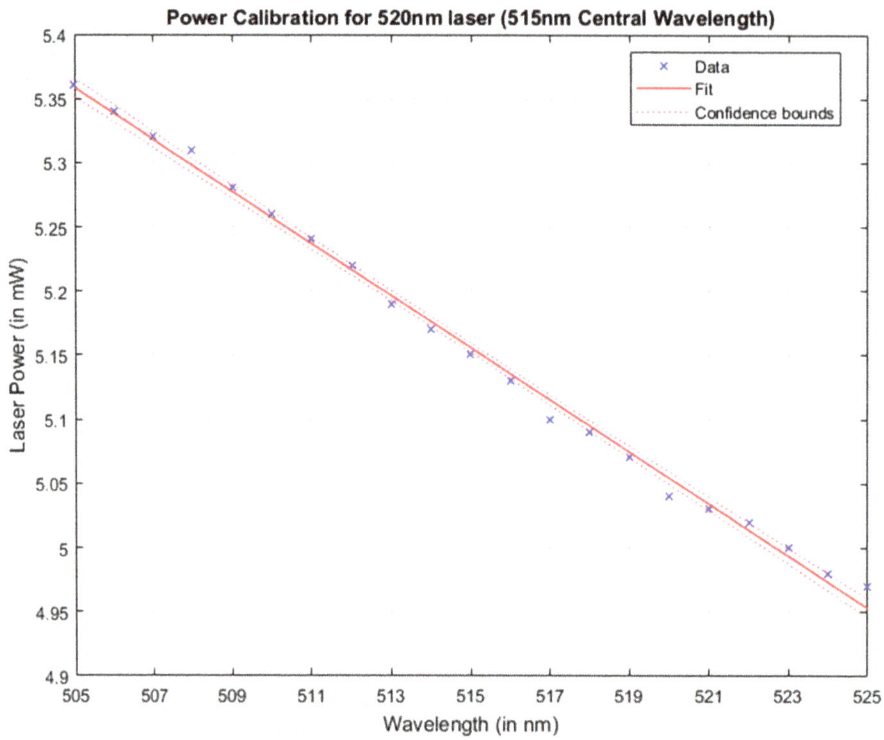

Figure 4.3. Laser power measured at $\lambda = \lambda_{CW} \pm 10$ nm, where $\lambda_{CW} = 515$ nm.

$$OD = \log_{10}(\frac{1}{T}), \quad \text{or} \quad T = 10^{-OD} \tag{4.1}$$

Although the filters are labelled according to their optical densities, in experimental conditions, their transmission percentages may vary to a certain degree. This variation is not a manufacturing defect, but is merely an implication of different operating conditions and incident laser's wavelength. For example, the transmission percentages listed on the Thorlabs Inc. website for the NE2xxB series of filters are listed at 633 nm, while we are using a 515 nm laser. Although files with individual wavelengths are also given, it is good practice to reaffirm the calibration with the applied equipment.

1. To accurately understand the **transmission of neutral density filters** at the experiment-specific wavelength, calibration of these filters is necessary. In order to calibrate them, we use the same experimental arrangement as that of the laser power measurement, adding in a mount to hold the filter in place, perpendicular to the beam's direction[4] (figure 4.2(ii)).

[4] The perpendicular orientation of the filter is important as the optical density of the filter could change due to a longer path length inside the filter when oriented at an angle to the beam's direction.

	A	B	C	D	E	F	G
1	**520nm Laser [515nm (Central Wavelength)]**						
2	NDF	Laser Intensity (mW)		T%	T%	Notes	
3	(OD)	Sans NDF	With NDF	(Given)	(Meas.)	• The laser is oriented towards the power meter, with the NDF in place.	
4	0.1	5.15	4.14	79.43282	80.38835		
5	0.2	5.15	3.02	63.09573	58.64078		
6	0.3	5.15	2.78	50.11872	53.98058		
7	0.4	5.15	1.99	39.81072	38.64078		
8	0.5	5.15	1.43	31.62278	27.76699		
9	0.6	5.15	1.18	25.11886	22.91262		
10	1	5.15	0.46	10	8.932039		
11	2	5.15	0.052	1	1.009709		
12	3	5.15	0.00231	0.1	0.044854		
13	4	5.15	0.00023	0.01	0.004466		

Figure 4.4. Illustration of transmission data recording in the Excel file when using the neutral density filters; Provided with the computer program on the MATLAB Community [1].

2. The laser power is measured without a neutral density filter in place to reaffirm the power of the beam.
3. The filters are placed individually, in the order of increasing optical density for each measurement. With this step, we can calibrate the individual filters along with their transmission percentages. During this experimental methodology, we use the NE2xxB series of filters, fabricated by Thorlabs Inc. The optical densities range from 0.1 (≈ 80% transmission) to 4 (0.01% transmission).
4. The laser power is reduced by the neutral density filter. Additional losses due to glass-related absorption at each surface of the filter are also accounted for in the measured intensity by the power meter. The reduced transmission is recorded (as illustrated in figure 4.4). The reduced transmission is used for **constructing the beam profile** by relating the reduced incident beam intensities on the sensor to the unobstructed beam profile.
5. The neutral density filters serve the purpose of reducing the overall intensity incident on the sensor. As the filter's absorption is uniform across the beam's profile, it proportionally absorbs the light energy, making the extrapolation of the beam profile possible by using the reduced intensity and the filter's calibration.
6. The transmission information is logged and plotted against the standard values (derived from equation (4.1)) to measure and estimate the error and plotted with error bars using (function ***errorbar()***) MATLAB (figure 4.5).

4.1.3 Imaging sensor calibration and application to laser diagnostics

At this stage of the calibration, the laser operator is in possession of a power-calibrated laser and a transmission-calibrated set of neutral density filters with

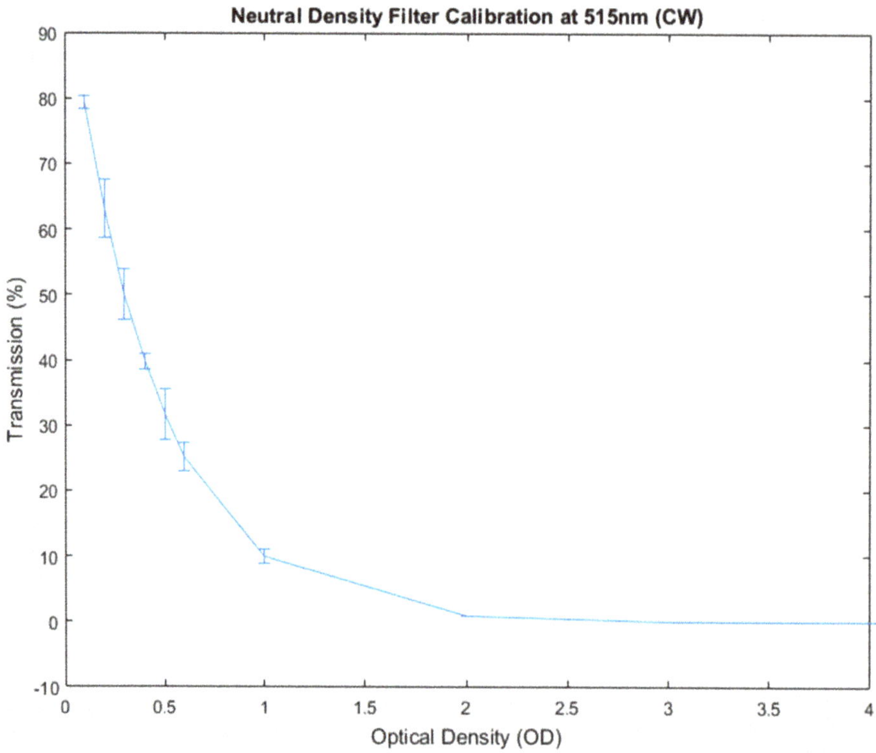

Figure 4.5. Transmission measurements at $\lambda = \lambda_{CW}$.

varying optical densities. The laser's power is only one dimension of the diagnostics this method is intended to provide. The distribution of power and shape of the beam are of a much higher significance, although it is derived from the knowledge of the calibration.

The imaging sensor used in this case is the DCC1545M monochrome CMOS imager, manufactured by Thorlabs Inc. This is a very compact and inexpensive CMOS-based camera with a small footprint, allowing it to be used in conditions where space is a constraint. The imaging system functions with an electronic rolling shutter, a feature that must be acknowledged when this method of laser diagnostics is applied to pulsed lasers where the nature of the shutter and the exposure time are vital considerations. If the timing of the imaging system is not synchronised with the pulsed laser system, the beam will not be captured appropriately, leading to a result where only a portion of the laser pulse was incident on the sensor at the time of capture.

1. For calibrating the imaging system, the highest priority is given to ensuring that no thermal or physical damage is induced on the sensor due to the direct incident of the laser light. The neutral density filters serve this purpose by reducing the intensity of the laser light by absorbing according to their optical densities.

(a) The experimental procedure builds on the previous two arrangements for calibrating the laser and the neutral density filters, while replacing the laser power meter with the CMOS imaging system (figure 4.2(iii)). The alignment of the laser and the imaging system can be checked by mounting the imaging system with the safety cover of the imaging system in place. When using a higher powered laser, a neutral density filter can also be placed to avoid damage to the plastic material used for the sensor covers.
(b) The imaging system is operated through a USB connection with a computer, using the ThorCam software package provided by the manufacturer. For convenience purposes, the exposure time is set to 'Auto' and the digital gain is set to '0'. The gain at zero restricts the software from automatically adjusting the image values to enhance the image, as opposed to capturing the true incidence.
(c) The software package allows two options—snapshot and video formats, for capturing the incident light energy. For **continuous wave lasers**, the snapshot mode can be used to capture the beam and construct the intensity and the shape of the laser beam, identifying the peaks and distribution of the laser power. The intensity can be estimated from the knowledge of the laser power transmitting through the neutral density filter and the beam's area (equation (4.2)). For **pulsed lasers**, the better method would be to use the video mode, with the frame rate clocked to its highest to estimate the incidence pattern, following which the intensity estimation can be applied over the series of images (a video capture is after all a series of images) to estimate the beam profile. Through this estimation, the frame rate of the camera and the exposure time for capture can be quantified to operate in snapshot mode[5].

$$\text{Intensity, } I = \frac{\text{Power (W)}}{\text{Area (m}^2)} \tag{4.2}$$

2. Prior to calculating the intensity and beam profile of the laser, the linearity of the sensor must be checked. Although this step does not need to be repeated as often as the rest of the calibration procedure, it is good practice as it helps in quantifying the performance of the imaging sensor. As with any equipment, the imaging sensor is also vulnerable to degradation over time and exposure, reducing the efficiency of the imaging sensor in converting light energy to electrical energy (quantum efficiency). It is vital to monitor the quantum efficiency of the sensors as there is the possibility of under estimation of the incident light when converted to digital values. Over time, with repetition of

[5] However, we would insist that the video operating mode be used for pulsed lasers and the beam profile be extracted in post-processing.

Table 4.1. Linearity check using multiple neutral density filters and measurements from the imaging sensor.

ND filter(s) used	Transmission (mW) × 10 000	Pixel count	Highest pixel value
OD4	2.304	11 604 113	193
OD4 + OD0.1	1.852 42	9 535 301	154
OD4 + OD0.2	1.313 28	8 161 079	119
OD4 + OD0.1 + OD0.2	1.055 88	7027837	112
OD4 + OD0.3	1.244 16	6 812 235	98
OD4 + OD0.4	0.898 56	5 267 852	80

this measurement, the performance of the sensor can be considered during measurement and experimental results can be duly corrected and/or adjusted.

(a) The linearity check is performed by monitoring the highest pixel value and total pixel count with varying incidence on the imaging sensor. The first time this measurement is performed, it establishes the benchmark of the performance of the sensor in the experimental conditions. The experimental arrangement is illustrated in figure 4.2(iii). The first image is collected with the highest pixel value achieved before saturation of the sensor. In this case, using the CPS520 laser and the filter with optical density 4 (0.01% transmission), the highest pixel value achieved is 193 (8-bit sensor).

(b) With the OD4 filter (NE240B) in place, we add additional filters to reduce the transmission to the imaging sensor. In the order, the optical density filter combinations are placed in the path of laser beam and their respective pixel counts are measured. Table 4.1 illustrates this information. For ease of discussion and presentation, the transmitted power of the lasers has been multiplied to reduce the number of decimal values. However, the original values prior to multiplication were used to plot the linearity plots.

(c) The data in table 4.1 was used to plot the linearity of the sensor's acquisition, presenting the plots (figures 4.6 and 4.7) using the ***fitlm()*** function in MATLAB.

3. With the information of the calibration and the capture of the beam profile, the intensity of the laser beam and its distribution can be calculated.

(a) Using a surface plot, ***surfc()***, in MATLAB, the beam profile can be visualised as a function of pixel value, and the horizontal and vertical pixels of the sensor. This provides the user with information regarding the orientation of the beam, with respect to the imaging sensor and the beam's shape. However, the intensity values at each point are not known at this stage. The total power incident on the sensor is known due to the laser power measured by the power meter for (set of) given neutral density filters (figure 4.2; table 4.1). The beam's footprint can be calculated from the number of pixels illuminated by the laser (any

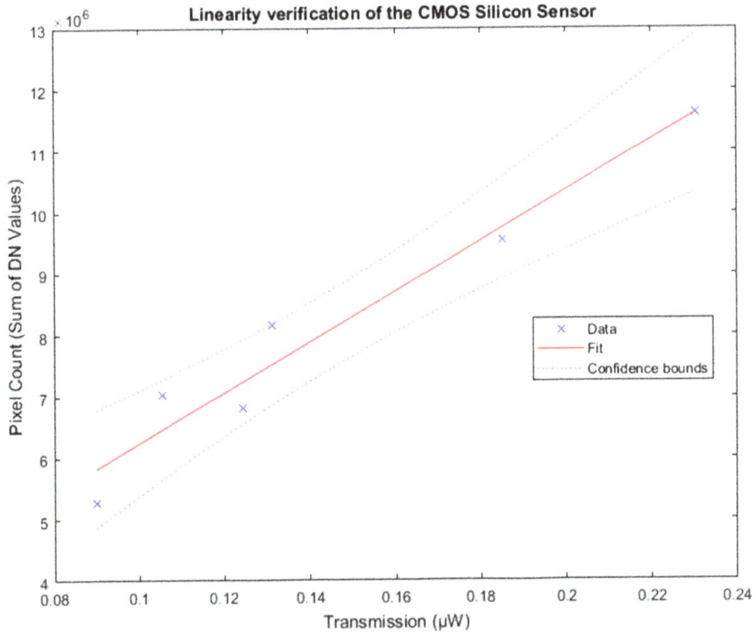

Figure 4.6. Linearity verification of the sensor's response utilising all the incident light on the sensor through the total pixel count of the image.

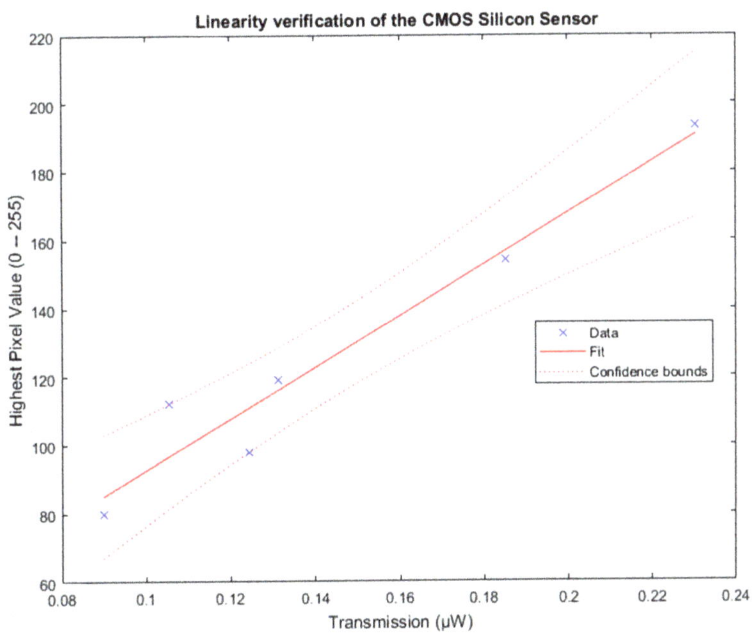

Figure 4.7. Linearity verification of the sensor's response utilising the single highest pixel value of the image.

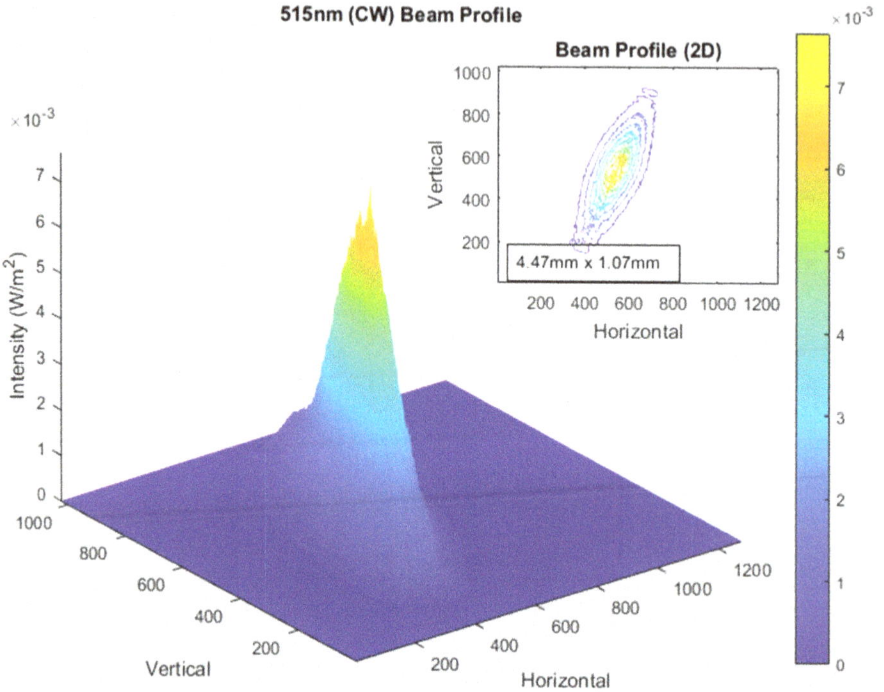

Figure 4.8. Calibrated intensity incident on the CMOS sensor, illustrating the beam profile and beam dimensions.

non-zero pixel, owing to the dark room experimental set-up) multiplied by the area of each pixel ($(5.2\ \mu m)^2$).

(b) Using the acquired image information, multiplied by the conversion factor (Pixel value × ((Laser power/Area)/Total count))[6], the laser intensity measurement is complete, but at the reduced transmission. This reduced transmission is due to the use of the neutral density filters. This should be accounted for and duly multiplied to obtain the beam profile and intensity distribution of the unobstructed laser.

As illustrated in figure 4.8, the intensity of the laser and the beam profile, including the footprint measures. The inset image shows the distribution with respect to the footprint, also presenting the measure of the footprint (obtained manually using the measuring tool of the provided software).

To reiterate for the benefit of the reader, the method illustrated above is available with the relevant code, templates and sample result images is available on the MATLAB Community page [1]. Alternatively, the page can be obtained by

[6] where Total count = Σ(Pixel value).

searching on the File Exchange page using the search keyword—**Laser, Filters and CMOS-Imager Calibration**.

Reference

[1] Kallepalli A 2017 Laser, Filters and CMOS-Imager Calibration (MATLAB File Exchange Computer Program Download)

IOP Publishing

Laser Beam Profiling: Cost-Effective Solutions

Akhil Kallepalli and David B James

Chapter 5

Potential sources of error

Abstract

Like any method of experimentation, there is always a possibility of inducing error into the process of calibrating the imaging system for use in laser diagnostics. In this chapter, we discuss a few possible sources of error and how to correct them.

5.1 Equipment alignment

One of the most common sources of error when using a laser source is the alignment of the equipment. The high degree of directionality allows the laser to maintain its intensity over a relatively longer duration and distance (in comparison to lamp sources). This property of the laser makes alignment of the equipment much more important, as variations of a few millimetres would result in highly reduced measurements. The alignment process when using lasers of visible wavelengths is comparatively simpler as opposed to infrared and ultraviolet lasers. Therefore, a low powered laser (green or red wavelengths) must always be used to confirm the alignment of the equipment before using these lasers. Alternatively, graded mounts placed on horizontal rails can also be used to align the laser.

Alignment is also important in the implementation of neutral density filters and the imaging sensor. The optical density of the filter is subject to change if the orientation of the filter is not perpendicular to the beam's direction. An inclined filter increases the path length of the light (more travel time and length inside the filter), changing the degree of transmission. Using rail-mount arrangements and other aligning tools, the filter should be placed appropriately in the path of the laser beam. The same degree of importance also applies to the sensor's alignment, as an inclined sensor is subject to misinterpreting the laser beam's orientation and shape. A round beam shape could be imaged as an elliptical footprint when the sensor is not aligned appropriately with respect to the laser.

doi:10.1088/978-0-7503-3835-6ch5

5.2 Sensor saturation

When light is incident on the imaging surface, the light energy is converted into an electrical signal and transmitted to the system for conversion into pixel value. In some cases, the light energy is higher than the conversion that the sensor can achieve. For example, an 8-bit sensor has the capability of reading the incident photons and presenting output pixel values between 0 (pure black) and 255 (pure white). In most cases, any value higher than 255 is also designated as 255 without providing information of how much higher. This process of obtaining the highest pixel value is called saturation. Saturation presents a very basic problem—if the pixel value is beyond 255 and there are other pixels below 255, it is almost impossible to compare these two values. Not only does this degrade the quality of the images obtained by the sensor due to white patches in the scene, it also induces physical damage on the silicon sensor over time, leading to the creation of dead pixels, i.e. pixels which do not convert the light energy to electrical energy (figure 5.1, figure 5.2).

Therefore, at every stage, the signal reaching the sensor must be lower than the highest value the sensor can achieve, in this case the DCC1545M is an 8-bit sensor. As in this method, when using neutral density filters to reduce the signal intensity reaching the imaging sensor, the key aspect considered to begin image capture was

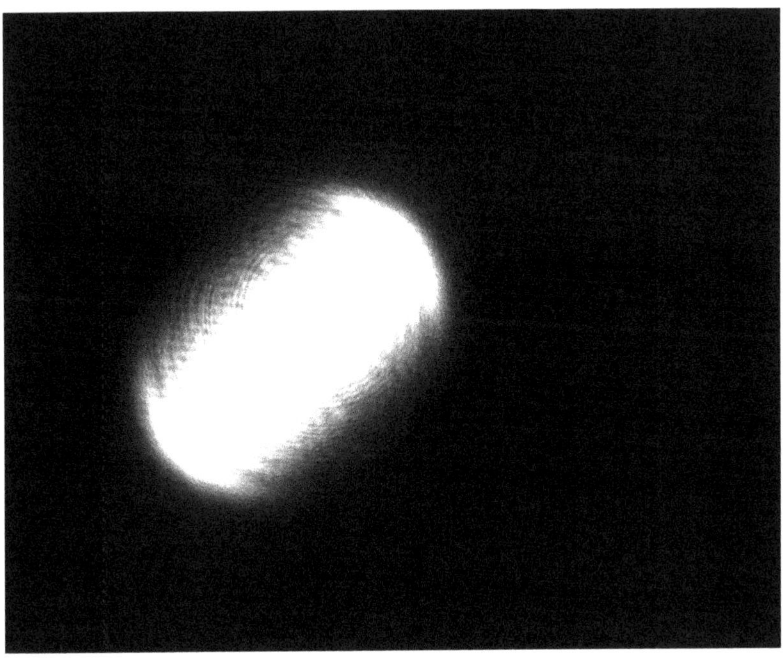

Figure 5.1. The imaging sensor can be saturated if the beam is not attenuated through the correct filters. In extreme cases, this can result in damaging the sensor permanently and therefore the saturation of the sensor must be avoided. If a sensor can store 8-bit data, then the highest pixel value measurable is 255. The sensor capture of saturation is given here. The experimental set-up is as given in figure 4.2.

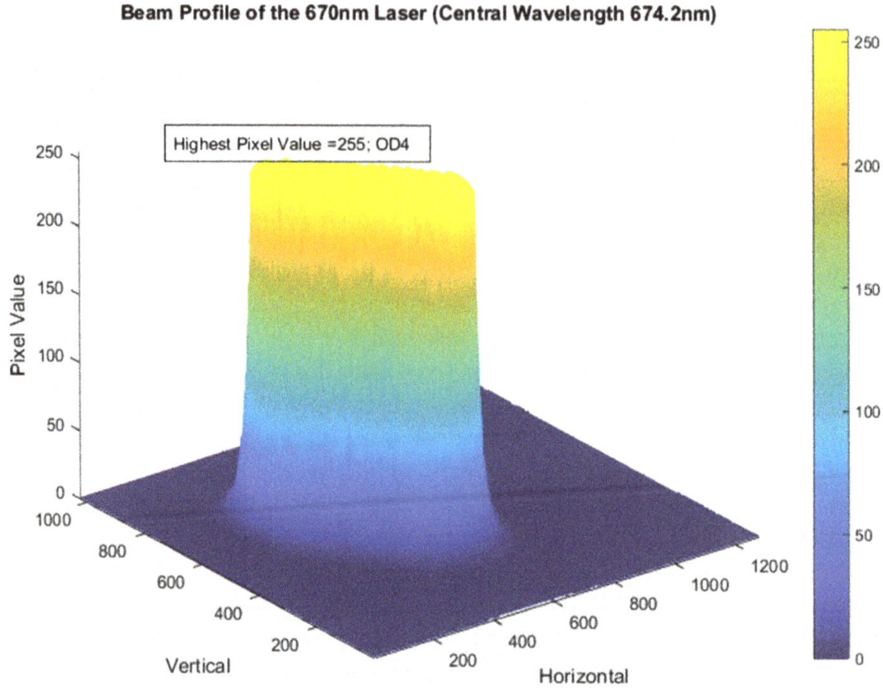

Figure 5.2. The laser beam profile of the capture shown in Figure 5.1 is calculated using the methodology in this text. The highest pixel value of 255 is measured and no usable information about the profile can be recorded due to saturation.

ensuring that the sensor was not saturated[1]. Similarly, it is very important to begin the laser alignment with the imaging sensor with a filter of low transmission and assessing the beam captured on the sensor. With due care of not exposing the sensor to the direct laser beam, the optical density in the path of the laser beam must be either increased or decreased until the highest acceptable pixel value is achieved, without saturation.

5.3 Neutral density filters

Neutral density filters usually operate in two modes—reflective and absorptive. While reflective filters have metallic coatings on their surface to achieve the reduced transmissions, absorptive filters are manufactured from Schott glass substrates which have spectrally flat absorption coefficients in the visible region of the electromagnetic spectrum. Thorlabs Inc. fabricates different optical densities by varying the thickness and type of glass, like the set of filters used in this literature (NE2xxB series, where xx refers to the optical density of the filter).

[1] When the OD4 filter is placed in the path of the CPS520 laser's beam, the highest pixel value achieved on the imaging sensor of the DCC1545M camera was 193.

Neutral density filters are vulnerable to degradation of performance over time; where thermal impacts and bleaching of the absorption properties of the glass substrate are widely seen. The repetition of the calibration of the filters using a laser of known power and a power meter (figure 4.2(ii)) at regular intervals of usage is good practice. Although the degradation may not be sudden, a reduction in performance could gradually increase the implications on the target, i.e. imaging sensor or any other biological or physical target depending on the application of the filter.

5.4 Dead pixels

Over time and regular usage and/or damage, as with any equipment, the sensitivity of the pixels in the pixel array of the sensor reduces. When pixels are continuously affected by usage and/or over-exposure, the pixels eventually lose their ability to convert the incident photons to voltage. The best method of spotting these is to remove the lens and direct the imaging sensor to a bright surface (like a wall or a whiteboard) or a diffused light source. The dead pixels manifest as black pixels in the resultant image capture. Following this, the pixel(s) can be corrected in post-processing or removed altogether as the location of these dead pixels will be known. A common correction method is averaging the value of the surrounding pixels to interpolate the pixel value of the dead pixel.

5.5 Laser stability

The laser emission is also known to vary with time, usage and experimental conditions. Manufacturers are known to calibrate lasers and their emissions prior to shipping, but at certain calibration conditions. This includes temperature, humidity, method and detector specifics. Once the laser is in the possession of the end user, it is possible that the laser could behave differently. Therefore, calibration of the laser is important. The calibration process discussed in this literature (figures 4.2(i)–(iii)) must be performed at the first instance of usage. This calibration provides a benchmark for the stimulated emission obtained from the diode module against which all future measurements can be compared. As with any electrical equipment, the performance of the laser will deteriorate with time and a measurement of this change should be factored in to all applications of the laser.

5.6 Stray light and background correction

The calibration of an imaging system or a light source is best done in a dark room experimental arrangement, as the dark room provides the truest measurement of incidence on the target. However, realising a true dark room is relatively difficult as there are always other sources of light in the vicinity. These could be from under doors, gaps in the window blinds and/or emergency lights on the ceiling. There could also be reflective surfaces in the room that could influence the experiment. This stray light adds into the measurement by the detector.

Any additional signal in the experiment must be measured prior to using the equipment for data acquisition. For example, in this experiment, prior to measuring the laser light's power with the power meter, we measure the incident light on the detector of the power meter with the laser off. The power measured of the background was under 1 μW (10^{-6}). While acquiring milli-Watts (10^{-3}) of power from the laser to the hundredth decimal, a background signal of under 1 μW can be disregarded. However, this step is important to quantify the background signal.

Similarly, the background signal is acquired from the imaging sensor with the laser turned off, in the dark room. This is done to quantify the incident light energy, within the sensitivity and field of view of the sensor. This background signal can then be subtracted from the image to remove its effect. In the case of this experiment, the background signal was considered from the corner of the image, i.e. if the beam is centrally located, then the pixels at the corners signify the background noise. This is subtracted from the image file to make the corner pixel value zero and accordingly adjust the value of the other pixels.

IOP Publishing

Laser Beam Profiling: Cost-Effective Solutions

Akhil Kallepalli and David B James

Chapter 6

Conclusions and final remarks

The primary purpose of this literature is to unify the applications of multiple instruments for simple and robust measurement of the laser emissions. A single CMOS or CCD-based imaging sensor has been shown to be applicable for both power measurements and laser diagnostics; i.e. measuring the beam's profile and the distribution of intensity. This assists in establishing the beam's footprint at the experimental distance and the characteristics of the incidence. Although 'laser diagnostics' encompasses a wide array of measurements, this literature focuses on two of the key aspects of the laser—beam profile and intensity distribution. The beam profile allows the visualisation of the shape of the beam and monitoring the changes over time with regards to its geometry and the presence of hot spots.

Traditionally, laser power measurements are done using power meters that provide a one-dimensional estimation of the power of the laser (Watts). This remains to be the most dependable method for power measurements and is, therefore, used to measure the laser's emission under experimental conditions. Once the laser's output has been calibrated, neutral density filters are calibrated for their transmission. Their transmission is a key parameter, as the purpose of the filter is to block a certain percentage of the laser's energy from reaching the sensor and damaging it.

Once the laser's power and the transmission of the filters are known, we use the imaging sensor to capture the incidence of the laser. As recommended, the imaging system is best operated in snapshot mode for continuous wave lasers while it is best suited to operate in video mode when profiling pulsed lasers. The incident energy is captured and and processed on MATLAB. The method establishes a benchmark of the laser's performance at the first use or when calibrated for the first time. Thereafter, regular repetitions of the procedure can assist in monitoring the performance of the beam over time.

Through the course of this literature, we have discussed some key aspects of today's optical technologies. Lasers, a unique type of light source, deliver high

doi:10.1088/978-0-7503-3835-6ch6

intensities of coherent light with a strong directionality in comparison to diffuse lamp sources. Their calibration is one of the key steps prior to their usage in applications as diverse as biomedical imaging to civil engineering. We have also discussed the two major types of imaging sensors with illustrations of their sensor and/or pixel design to understand their functioning.

www.ingramcontent.com/pod-product-compliance
Ingram Content Group UK Ltd.
Pitfield, Milton Keynes, MK11 3LW, UK
UKHW051847100926
8556IPUK00005B/507